Fundamentals of Engineering Programming with C and Fortran

Fundamentals of Engineering Programming with C and Fortran is a beginner's guide to problem solving with computers that shows how to prototype a program quickly for a particular engineering application. The book's side-by-side coverage of C and Fortran, the predominant computer languages in engineering, is unique. It emphasizes the importance of developing programming skills in C while carefully presenting the importance of maintaining a good reading knowledge of Fortran.

Beginning with a brief description of computer architecture, the book then covers the fundamentals of computer programming for problem solving. Separate chapters are devoted to data types and operators, control flow, type conversion, arrays, and file operations. The final chapter contains case studies designed to illustrate particular elements of modeling and visualization. Also included are five appendixes covering C and Fortran language summaries and other useful topics.

The author has provided many homework problems and program listings. This concise and accessible book is useful either as a text for introductory-level undergraduate courses on engineering programming or as a self-study guide for practicing engineers.

Harley Myler is a professor of electrical and computer engineering at the University of Central Florida in Orlando. A senior member of the IEEE and a member of SPIE, he earned his Ph.D. and M.Sc. at New Mexico State University. He is the author of two other books: *Computer Imaging Recipes in C* (1993) and *The Pocket Handbook of Image Processing Algorithms in C* (1993), both published by Prentice-Hall.

Fundamentals of Engineering Programming with C and Fortran

Harley R. Myler

CAMBRIDGE
UNIVERSITY PRESS

PUBLISHED BY THE PRESS SYNDICATE OF THE UNIVERSITY OF CAMBRIDGE
The Pitt Building, Trumpington Street, Cambridge CB2 1RP, United Kingdom

CAMBRIDGE UNIVERSITY PRESS
The Edinburgh Building, Cambridge CB2 2RU, UK   http://www.cup.cam.ac.uk
40 West 20th Street, New York, NY 10011-4211, USA   http://www.cup.org
10 Stamford Road, Oakleigh, Melbourne 3166, Australia

First published 1998

Typeset in Stone Serif 9.5/14 pt. and Antique Olive in LaTeX [TB]

Library of Congress Cataloging in Publication data
Myler, Harley R., 1953–
    Fundamentals of engineering programming with C and Fortran / Harley
R. Myler.
        p.   cm.
    Includes bibliographical references and index.
    ISBN 0 521 62063 5 hardback
    ISBN 0 521 62950 0 paperback
    1. C (Computer program language)   2. FORTRAN (Computer program
language)   3. Engineering – Data processing.   I. Title.
    QA76.73.C15M93   1998
    005.13 – dc21   97-43343
                                                                          CIP

A catalog record for this book is available
from the British Library

ISBN 0 521 62063 5 hardback
ISBN 0 521 62950 0 paperback

Transferred to digital printing 2004

To my son, Logan

Contents

Contents

Contents

Preface

This text is intended as an entry-level treatment of engineering problem-solving and programming using C and Fortran, the predominant computer languages of engineering. Although C is presented as the language of choice for program development, a reading knowledge of Fortran (77) is emphasized. The text assumes that any Fortran code encountered by the reader is operational and debugged; hence, an emphasis is placed on a reading knowledge of this language. Fundamental approaches to engineering problem-solving using the computer are developed, and appendixes that serve as ready reference for both languages are included. A basic premise of this book is that the engineer, regardless of discipline, is more interested in fast program prototyping and accurate data outputs than in program elegance or structure. The novice engineering programmer is concerned principally with modeling physical systems or phenomena and processing accurate data pertaining to those systems or phenomena. These are basic tenets of engineering programming that are subscribed to in this book.

In the introductory chapter, an understanding of basic computer architecture using the von Neumann model is developed as a register–ALU–memory (Arithmetic Logic Unit) transfer system. This concept is then integrated into an explanation of Tannenbaum's virtual machine hierarchy to illustrate the multiple levels of translation and interpretation that exist in modern computers. The relationship of programming languages to this hierarchy is then explained through diagrams and illustrations to enable the reader to develop a strong mental picture of computer function through language. This aspect of programming is often ignored by other texts; however, the critical dependence of data accuracy on the architecture of the implementing platform, particularly with respect to variable typing, demands that these concepts be understood by the engineering programmer. Discussions of computer architecture in this text are at a browsing level so that engineers from disciplines other than computing can feel comfortable with the explanations. In spite of this, electrical and computer engineering students should find the discussions an interesting introduction to subjects that they will explore in greater detail later in their training.

In Chapter 2 the edit–compile–run cycle is presented as the primary method of program development. Please note that no emphasis is made on any particular compiler or development system – these choices are left to the reader or instructor to make. Additionally, the text does not emphasize a particular computer platform owing to the wide range of machines encountered in engineering practice. Techniques for algorithm development using flowcharts and pseudocode are discussed, and these vehicles of algorithm representation are used throughout the text. This book is not intended to be a software engineering text, and thus only rudimentary concepts from this area are discussed.

Chapter 3 introduces types, operators, and expressions along with console input–output (I–O) methods. Examples of programs that simply process arithmetic and algebraic expressions are shown to introduce the reader to actual program coding and gross data processing. Chapter 4 discusses the use of fundamental language constructs for control flow in program decision making and loop construction. All of these topics are presented with engineering problem examples. Chapter 5 explores data type conversion as a prelude to the writing and use of functions. These concepts lead into the scope of variable activity within the program. These topics are typically introduced sooner in other presentations; however, most program errors are related to bad typing or type mismatch followed by errors of function definition and scope. Because a C program begins with the definition of the *main* function, expansion of this aspect of the language follows cleanly when functions are introduced late in the text. Chapter 6 discusses structures and pointers and their use in creating and working with array variables. The C language *union* and *typedef* are not discussed. Chapter 7 is a short introduction to file operations to include both low and high level I–O. Chapter 8 completes the book with case studies of two complex programs.

The book is self-contained and useful as a self-study tutorial or as a text for a one-semester introductory engineering programming course for students with no prior computer programming experience in either C or Fortran. Each section covered includes student exercises and programming examples. A set of instructor materials is available that includes overhead transparency masters and quiz and examination problems. The text was developed and tested over nine semesters at the University of Central Florida in our EGN3210 Engineering Analysis and Computation course. This course is required as a prerequisite for our numerical methods course for undergraduate students of all engineering disciplines who have had no prior computer programming instruction.

Although responsibility for this work is uniquely mine, I would like to thank all of the students personally who suffered through numerous editions of the text starting with overhead projector notes and culminating with rough drafts of the manuscript. May you always get the correct answers from your programs.

Orlando, Florida   Harley R. Myler
May 1997

1 Introduction

Some dictionaries define an engineer as a *builder of engines*, and it is relatively easy to classify engineering fields using this definition. Purists will insist that modern engineers rarely dirty their hands actually building anything; however, we can, without loss of the thread being developed here, include the *design* of engines within the definition. For example, many electrical engineers build (design) electrical engines such as motors and generators, and automotive engineers often build internal combustion engines. We can abstract the concept of engine to include machines in general as well as complex machines such as robots and vehicles. To further the abstraction, we can include systems that transfer or convert matter or energy from one state to another under the umbrella of machine design. Examples of such systems are water treatment facilities, the domain of civil engineers, or automated manufacturing facilities that attract the attention of industrial engineers. A computer is nothing more than an information processing engine. Now the material to be processed has been taken to the highest level of abstraction, the symbolic level.

The complexity of the world we live in, with the astonishingly high rate of information exchange and shrinking global barriers, demands that engineers utilize and command information processing systems. Computers are at the core of all nonbiological information processing systems, and they process the information that they are given with strict attention to detail. The level of detail is extreme, and the process by which we specify the details of the task that we wish the computer to perform is called programming. It is essential that the modern engineer, independent of engineering discipline, learn how to operate and program the computer.

1

If gasoline that has lost combustibility from long-term storage, or that has been corrupted by moisture, is used in an internal combustion engine, it would be no surprise to observe inadequate engine performance – if the engine will run at all. Why then should a computer be expected to process bad data? Further, if an engine is poorly designed for the fuel that it must use, should one still expect optimal performance? Why then expect good performance from a computer that is running a poorly written program? The engineer can apply the same principles of engineering design that are used to build machines to the construction of computer programs. Doing this, the engineer can develop programs that are effective and efficient in processing data for any engineering application.

This chapter begins by outlining a brief history of computing and discusses, as simply and illustratively as possible, the fundamentals of computer design and architecture. This material, although trivial to the experienced computer engineer, is often overlooked or ignored in the training of noncomputer specialists. To evaluate the output of a computer adequately, regardless of engineering purpose, it is important to understand these fundamentals.

1.1   History of Computers

The first computer that most humans encounter is the digits of their hands. When human civilization began to process numerical concepts using fingers to count on is unknown, but it was probably shortly after we discovered bartering. Not surprisingly, commerce has done as much to advance computing as science and engineering have. A case in point is that the acronym IBM stands for International *Business* Machines. Long before the formation of the IBM company, however, an English professor of mathematics named Charles Babbage (1792–1871) formulated the concept of a numerical computing engine. His first machine, the **difference engine**, was built in the early 1800s. This machine was designed to run a single program that computed tables of numbers for use in ship navigation – a subject of great interest to shipping merchants of the time. The name *difference engine* came from the method of finite differences that it used to compute the tables. The second machine that Babbage designed, the **analytical engine**, was a substantial improvement over the difference engine in that it could be programmed using punched cards, thus allowing any mathematical operation to be performed. Babbage never finished the analytical engine; however, the British Museum commissioned the construction of a machine from his original plans that is now on permanent display in their collection. In spite of his failure to build an analytical engine, Babbage hired Ada Lovelace, daughter of the English poet George Gordon, to write software for the machine. Thus, Babbage not only established the first computer programmer, but he also demonstrated the modern-day practice of software development for an architecture occurring in parallel with hardware development. It should be noted that the ADA® programming language developed by the U.S. Department of Defense was named in Ada Lovelace's honor.

Babbage's computing engines were mechanical devices, and it was nearly a hundred years later that Konrad Zuse (1910–1995), a German engineer, built a calculating machine called the **Z1** using electromagnetic relays. Zuse was planning to add programmability to his machines when the Allied bombing of Berlin during World War II brought his work to a halt. Ironically, war accelerated the need for fast computing machines in two ways. First, the British needed a computer to run the decoding procedures developed by Alan Turing (1912–1954), a mathematician, to break the codes generated by the German Enigma message encryption machines. Secondly, the Americans needed a computer to calculate trajectory data rapidly for the artillery. In reponse to these needs, the British developed **Collossus**, the world's first electronic computer, which was successful in breaking the Enigma codes using a program developed from Turing's work, which was kept a closely guarded secret for many years after the war. Many historians credit the cracking of the Enigma codes as a primary contribution to the winning of the war by the Allied forces. An American machine, the Electronic Numerical Integrator and Computer (ENIAC) was completed in 1946 but was introduced too late to be of any use in the war effort. Nevertheless, the ENIAC machine formed the basis of the first commercial computers built by the Univac Corportaion. The ENIAC machine has been preserved as an historical item by the Army Research Laboratory (ARL). The ARL has established a World Wide Web page (`http://www.arl.mil`) that you may browse for further information to learn the history of ENIAC. After World War II, research into the design and construction of electronic computing machines accelerated and has not slowed, even to this day.

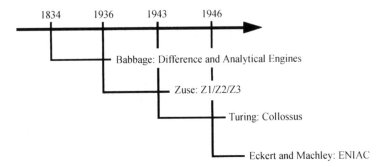

Figure 1.1 Early computer development timeline.

It should be noted that a major breakthrough in engineering came about from the invention of the slide rule, which is simply a mechanical analog computer. When hand-held calculators appeared in the late 1960s, their arrival marked the end of the usefulness of slide rules. Hand-held calculators will someday be replaced by palmtop computers and ultimately by communications devices that will link us with machines that understand our speech. All of these devices have evolved from the historical roots discussed in this section (see Figure 1.1) and, until a computer is built that can learn, will continue to require programming.

1.2 The von Neumann Machine Architecture

John von Neumann (1903–1957), a Hungarian-born mathematician who emigrated to the United States in 1930, first conceived the idea of the stored program computer, now known as the **von Neumann machine**. A **program** is formally defined as a sequence of instructions describing how to perform a task. For example, you could be *programmed* to make hamburgers a certain way by the following set of instructions:

BURGER CONSTRUCTION PROGRAM

1. Get bun and open it on counter.
2. Place all-meat patty on bottom piece of bun.
3. Place tomato slice on patty.
4. Place lettuce leaf on tomato.
5. Squirt special sauce on lettuce.

6.   Replace top of bun; burger is complete.
7.   Wrap burger in paper and place on warming tray.

Of course, the assumption is that the instructions make sense to you and that you can follow them. It is further assumed that the *data* in this program (the bun, patty, tomato slice, etc.) are available to you at execution time and that they are in the proper form. The program could be made more complex by specifying a *cooked* all-meat patty, but you assumed that, didn't you? Please don't be insulted. If this were a *computer* program, these would be important details to consider!

An **algorithm** is a formal term for a detailed set of instructions on how to perform a computation. At first glance it seems that there is no difference between an algorithm and a program. Algorithms are developed as a mathematical exercise, or general method, to achieve a computational result. A program consists of a set of instructions (to a machine) developed from the algorithm. A program is an algorithm, but an algorithm is only a program when it is specific to an implementation. The burger construction program is a set of instructions to a human cook. Likewise, a C or Fortran program, which we discuss in much detail later, consists of instructions to a computer.

Early machines had fixed algorithms performed by programs that were designed into the machine architecture, such as the computation of logarithm tables by the Babbage difference engine. The data were internal to the algorithm in that they started as a fixed value and were either incremented or calculated. Programmable machines allowed the algorithm processed by the machine to be changed, thus broadening the utility of the computer. In early electronic machines such as ENIAC and Colossus, programming was accomplished through a tedious method of changing control and data pathways manually by the use of wire jumpers and switches. Data entered the machine from punched cards or were inherent to the program. These machines were difficult to program and nearly impossible to **debug**, which means to find problems in the program. An interesting fact is that the term *computer bug* comes from an early machine in use by the U.S. Navy that stopped working one day. The problem was found to be a moth that had crawled into a relay and was caught between the contacts, preventing the proper operation of the part and retroactively, the program. From that moment, when a computer would not run properly, it was said that the program "had a bug in it." The

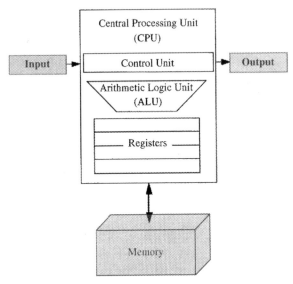

Figure 1.2   Von Neumann machine architecture.

coining of this term is attributed to Admiral Grace Hopper, an early pioneer in computer development.

The structure of a von Neumann machine allows both program statements and data to reside simultaneously in the memory in contrast to the early machines in which programming instructions were contained in a unit of the computer separate from the data. All modern computers are based, in part, on the concept of the stored program computer, or von Neumann machine. The von Neumann machine has five basic parts, as illustrated in Figure 1.2, that we collectively refer to as the **architecture** of the computer. It is important for the programmer to understand this simple yet powerful structure because it has a direct relationship to how we program the machine.

The *control unit* orchestrates the passage of data between the other units of the machine. It is the control unit that interprets the instructions of the program. When we direct a machine to do something, we are telling the control unit what we want done. Calculations and data manipulations occur in the *arithmetic logic unit* (ALU). Results of calculations performed in the ALU can be used by the control unit to redirect or change data pathways. In other words, if the result of a calculation is zero, we may want the computer to do one thing, and if the result is nonzero we may want the computer to do something else. It is this decision capability that makes the computer a powerful tool and distinguishes it from a calculator.

The *memory* is a storage unit for machine instructions and data. Think of the memory as a scratch pad. Written onto the pad are the program instructions followed by the control unit. Part of the pad is available for jotting down intermediate results or notes about how the calculations are proceeding. The *input* unit allows data to enter the machine from external sources, whereas the *output* unit allows the machine to display the results of its computations.

The control unit is constructed in such a way that it accepts binary data representing coded machine instructions when the machine is activated. These instructions are made available in the memory. The instructions are then fetched from memory and executed by the control unit until a halt instruction is encountered. The instructions can cause the control unit to input data, read data from memory, output data, write data to memory, or process data in the ALU. The way that the machine processes data, the range and format of data that the machine can handle, and the type of instructions that the machine can interpret are dependent solely on the machine architecture.

1.3 Binary Numbers

Computer data are stored and processed in **binary**, or base 2, form. We are familiar with the base 10 system primarily because we each typically have ten fingers! It should be no surprise that our numbering system is based on this count, or **radix**. To work, a number system requires a set of unique symbols, and the radix determines how many symbols are needed. In the base 10 system, we use the symbols 0 through 9, a total of ten symbols. As we count, when we reach the upper limit of the radix, we cross over to the next power of the radix to represent increasingly larger numbers. You have been doing this kind of counting for years and have memorized how to count to very large numbers in the base 10 system. For example, the number 403 is a shorthand for

$$\underline{4} \times 10^2 + \underline{0} \times 10^1 + \underline{3} \times 10^0$$

An electronic computer does not have ten digits to represent numbers with. Instead, it has available only the state of an electrical signal, which is either on or off, present or absent. Hence, a computer is restricted to a radix two, or **binary**, numbering system. Some people panic at the thought of having to learn the binary system. They say,

How can just 0 and 1 allow me to count to large numbers? The key is that counting systems are exponential; they increase in powers of the radix as the position of the significant digit (a fancy way of saying the digit we are working with) changes. The digit 4 in 403 has greater significance than the digit 3 because it is a factor of 100, whereas the 3 is a factor of 1. Now try to put the same reasoning to work to understand binary numbers.

The number 403 in binary is 11001001_2. The binary system is not as compact or efficient as the decimal system because the radix is only one-fifth as large. Nevertheless, we can represent very large numbers with the binary ranges found in modern computers. Note that we will use a subscript to indicate a radix of other than 10; otherwise, we might interpret the binary number above as 110,010,011! Expanding the binary version of 403 as we did above yields

$$\underline{1} \times 2^8 + \underline{1} \times 2^7 + \underline{0} \times 2^6 + \underline{0} \times 2^5 + \underline{1} \times 2^4 + \underline{0} \times 2^3 + \underline{0} \times 2^2 + \underline{1} \times 2^1 + \underline{1} \times 2^0$$

To simplify this expression, we have $256 + 128 + 16 + 2 + 1 = 403$. The system may appear alien to you because we are so accustomed to the decimal system. If you had spent your early years learning binary instead of decimal, you would quickly and easily interpret binary numbers on inspection – as von Neumann is reported to have been able to do! As it is, binary numbers are easy to use because the powers of two simply double as the exponents increase, $1 \to 2 \to 4 \to 8$, etc. We call the place, or power, in the decimal system a **digit**. In computing, we call each place in a binary number a *binary digit*, or **bit**, for short. The conversion of decimal numbers to binary is complicated because it involves repeated divisions of 2, but the conversion of binary to decimal is, as seen in the example above, very straightforward.

The size of binary numbers in computers varies according to the architecture and the computer languages used. As a result, several terms are used to describe binary numbers. A group of bits is called a **word**. Words can have varying lengths; however, 8-bit words have a special name, the **byte**. Occasionally one hears the term **nibble** (or nybble) for a 4-bit word, or half-byte. A byte of data can represent decimal numbers from 0 to 255, as shown in Table 1.1. The number of bits in a word tells you how many numbers it can represent: just take two to the power equal to the number of bits. Hence, a byte can represent $2^8 = 256$ numbers. The maximum number represented,

Table 1.1 *Data Byte Representation in Decimal and Binary.*

Decimal		Binary
0	→	00000000_2
1	→	00000001_2
2	→	00000010_2
⋮		⋮
126	→	01111110_2
127	→	01111111_2
128	→	10000000_2
⋮		⋮
254	→	11111110_2
255	→	11111111_2

however, will be one less to account for the zero at the beginning of the sequence: $0, 1, 2, \ldots, 254, 255$.

Computers generally express input and output data as decimal numbers as well as letters that correspond to written language. The internal representation, however, is binary. More extensive interpretation of binary data to express words and decimal numbers will be discussed in later chapters of this book. In all cases, the computer has specific and clearly defined mechanisms of interpretation that are very important to the engineer if data analysis pitfalls are to be avoided. It is for this reason that you must become familiar with the binary representation of numbers.

When we speak of very large numbers of bytes (such as are found in memory systems, disk drives, and communications channels), we use a set of abbreviations listed in Table 1.2. If we have 1,024 bytes, then we say we have 1 K bytes (pronounced one-kay bytes). An easy way to remember the exact value of the notation is to multiply the number of K bytes by 1,024. For example, 64 K is just $64 \times 1,024 = 65,536$. When we reach 1,048,576 bytes, we say *one megabyte*, and so on. Higher numbers follow the International System (SI) prefixes (giga, tera, etc.). Because this convention is also used when describing amounts of bits or words instead of bytes, be careful of the context.

Table 1.2 *Abbreviated Notation for Large Powers of 2.*

Power	Number	Notation
10	1,024	1 K
11	2,048	2 K
12	.4,096	4 K
13	8,192	8 K
14	16,384	16 K
15	32,768	32 K
16	65,536	64 K
17	131,072	128 K
18	262,144	256 K
19	524,288	512 K
20	1,048,576	1 M

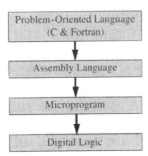

Figure 1.3 Virtual machine hierarchy.

1.4 Virtual Machine Hierarchy

Modern computers exhibit a structure that is useful in the study of how programming relates to real-world problems and to the architecture of the computer. This structure is called the *virtual machine hierarchy* and is illustrated in Figure 1.3. What we mean by a **virtual machine** is that at each level a machine is defined with all of the features of a von Neumann architecture (see Figure 1.2). Whether or not this machine exists as hardware or software is unimportant – we are only interested in the behavior of the machine at this point. At the bottom of the hierarchy is the **digital logic level**.

At this level, the electronic circuits that perform the logic necessary to generate computations are found. Recall that the computer works with binary information. An entire algebra is defined around binary quantities and is called **Boolean algebra** after the English mathematician Robert Boole (1815–1864). Using electronics that sense on and off conditions, this algebra is implemented as the fundamental control and computational structure of the modern digital computer.

Machine code is the term used to describe the binary coded instructions that are executed directly by the digital logic. The program that implements these instructions is known as a **microprogram**. Users do not have access to the microprogram, for the machine designers determine how many codes the processor will respond to as well as what the codes will do during execution of the program developed from them. These codes are called the **processor instruction set**, and they are very enigmatic to anyone but the machine designers. As a consequence, an **assembly language** is provided to simplify the programming of a processor at this level. Assembly languages are unique to a processor class, and manufacturers try to make the assembly codes of sequential processor models compatible with earlier processors in the series. Nevertheless, the assembly programs of one processor will not run on a processor outside the processor class. Two examples of this are the Motorola 68000 series processors (68000, 68010, 68020, 68030, and 68040) and the Intel 80×86 series (80286, 80386, 80486, 80586 – Pentium). Programs written in 68000 code will run on the 68040, and programs written in 80286 code will run on a Pentium (80586), but 68xxx code of any kind will not run on any of the 80×86 series processors. To put this difference into perspective, the Apple Macintosh uses Motorola processors, whereas the personal computer, or PC, uses Intel processors.

The assembly language program is *assembled* by an **assembler**, which is just a program for converting from assembly code to machine code. The code produced by the assembler is called an **object code** and must be linked to other codes to be useful. The linking process is accomplished by a **linker** or **loader** program. After linking, the program becomes an **application**, or user-oriented program, that performs a useful task. The application is what we are interested in programming or using.

At the highest level of the hierarchy, the **problem-oriented language level** may be used instead of, or in conjunction with, the assembler to produce an application. This is generally the preferred approach because the machine can be instructed in an easy-to-understand language. The **C** and **Fortran** computer languages are examples of problem-oriented languages. Fortran was designed to make the programming of mathematical formulas easy. The word Fortran is a conjunctive acronym for **For**mula **Tran**slation. The C language was developed to write **operating systems**. An operating system (OS) is a special application that manages the resources of a computer system. Because of the size and complexity of most operating systems, it is best to write them in assembly language for reasons of efficiency and size. However, a problem arises when one wishes to port, or transfer, an OS from one class of computer to another. If the OS is written in assembly language, then it must be rewritten for each computer it is to run on. This is always an expensive, laborious, and time-consuming task. The C computer language was originally developed to write operating systems and to allow the easy transfer of those operating systems from machine to machine. For this reason C is very similar to assembly language, particularly in the ability to manipulate memory. Because of the efficiency, speed, and simplicity of the C language, it has become a dominant player in engineering programming.

Problem-oriented languages are implemented by a **compiler**, which is a computer program that generates an application from a program written in the problem-oriented language. The compiler often includes the linking and loading functions that make application programming all the more simple. Most modern compilers are written in C, and major parts of C compilers are also written in C. At some point, of course, the recursion must end. Because the basic compiler functions are written in assembly language, the compiler becomes unique to a processor or processor class. Typically, the first compiler to become available after a new processor is designed is the C compiler. This facilitates the porting of software from one processor to another, and of all computer languages, C is one of the most portable.

As one moves up in the virtual machine hierarchy, there is an increase in abstraction from what is actually happening down at the digital logic level. Recall that the digital logic level is electronic, and the speed and efficiency of computation is restricted only by the technology used to build the hardware. For example, a fast 8-bit machine can multiply as quickly as a slow 16-bit machine. At the top of the hierarchy one rarely cares about the details of what is happening down below. Nevertheless, if errors are made at lower levels, they will propagate to the higher levels and, in some cases, be very difficult to detect. This fact was brought home when floating point mathematical errors were discovered in the early releases of the Intel Pentium processor. The errors were noticed by scientists using the processor to perform complex simulations and went unnoticed by the general public until the errors were discussed by the popular media.

1.5   Register–Memory–ALU Transfer System

Memory in a computer is described in terms of binary words, and a word can be any number of bits. Recall that the size of a word is strictly determined by the computer architecture. Also recall that a **byte** is a term given to an 8-bit word. The smallest unit of memory in a processor is called a **register**, which consists of a single word of very fast memory in the control unit. The number of registers and the number of bits that the registers can hold are dependent on the processor. For example, a 16-bit processor has 16-bit registers that can process 2 bytes of data at one time. Figure 1.4 shows how a register may be depicted graphically; the register is 8-bits (1-byte) wide and contains the number 195.

Registers are typically designated by letters, such as A, B, C, and so forth. They are used at the microprogram and assembly language levels and occasionally at the problem-oriented language level. Modern processors will have between sixteen and thirty-two registers available. Assembly language programs consist of instructions called **mnemonics**, and these programs specify actions to be taken by the processor at the register level. Most modern compilers, particularly C compilers, allow assembly language to be inserted into the

| 1 | 1 | 0 | 0 | 0 | 0 | 1 | 1 |

Figure 1.4   8-bit (one-byte) register with binary value $11000011_2 = 195$.

Address	Cell
0000000000	10101010
0000000001	00011000
0000000010	01001100
0000000011	00001111
0000000100	11011110
0000000101	00110101
0000000110	00111101
0000000111	11110111
⋮	⋮
1111111111	100100101

Figure 1.5 1 K
memory of 8-bit
(one-byte) cells.

high-level program when optimum speed or efficiency is desired. A
typical assembly language program might look as follows:

```
LOAD A, MEM1    ;Read data from memory to registers A and B
LOAD B, MEM2
ADD A,B         ;Add contents of A to B

LOAD C,A        ;put result in C
```

The main memory of a computer is called a random-access memory (**RAM**). The size of this memory is anywhere from 1 K to 128 M
bytes, depending on the size and sophistication of the computer system. The main memory is used for program and data storage, although secondary memory such as disk and tape play an important
role in program and data storage. The main memory is arranged as a
set of **cells** that contain the actual information being stored. These
cells are accessed by a binary word that identifies the cell **address**,
or location, in memory. Figure 1.5 shows the arrangement of a 1 K
(1,024 bytes) memory.

The memory of Figure 1.5 illustrates the difference between the
two binary numbers used in memory systems. In this case, a 10-bit
number is used to determine the address. The range of addresses is
then 0000000000_2 (0) to 1111111111_2 (1023), or 1,024 cells. Each

cell contains 8 bits of data or a binary number that can represent 0 through 255. Note how the addresses count up from zero, although the cells appear to contain random values. The addresses in memory are sequential like the address numbers on houses on a street. Each house will contain a different family or individual in the same way that each cell contains a different byte of binary data.

The control unit of the computer interprets the data in the cell as either an instruction to be followed or an item of data to be processed. How and when this takes place is beyond the scope of this book; however, it is important for the engineer to understand that binary data are moved from the main memory to and from the registers and also to and from input/output (I/O) devices. When variables are declared in the user program, they are represented by memory locations. The interpretation of the variables and the data they contain is dependent on the programming language used.

Numerical calculation and symbolic decision functions take place in the ALU under the direction of the control unit. For example, the control unit, using binary codes (instructions) fetched from memory, can command the ALU to add the values contained in two registers and place the result in a third register. The results may then be written to the main memory for access at a later time or outputted to a device. This structure is illustrated in the two-register system shown in Figure 1.6. The ALU is capable of addition and subtraction, and combinations of these operations lead to multiplication and division. The computer program specifies when and where data are transferred (the binary numbers) and what operations the ALU performs on the data. For example, the data in Register A can be transferred to an output device (such as a computer screen or printer), or new data can be brought in from an input device (such as a keyboard

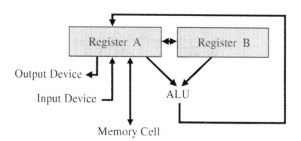

Figure 1.6 Register–memory–ALU transfer system.

or mouse). Because data can also be stored and retrieved from *memory*, the machine is capable of remembering sequences of complex calculations.

All digital computers possess this fundamental architecture, and you can readily see that it is a von Neumann machine. Computers differ only in the number of registers they possess, the amount of data (size of the binary word) that the registers can hold, the amount of memory that they can access, and the speed at which the transfer operations take place. In the programming techniques discussed in this book, we will replace the register with a variable and the ALU with an arithmetic or logical operator. Other than this, the basic concepts of the register–memory–ALU transfer will remain the same.

REVIEW WORDS

address
algorithm
analytical engine
application
architecture
arithmetic logic unit (ALU)
assembler
assembly language
byte
cells
compiler
control unit
difference engine
Fortran
high-level language
linker
loader
machine code
memory
mnemonics
nibble
object code
operating system (OS)

problem-oriented language
program
radix
register
von Neumann machine
word

EXERCISES

1. Convert the following binary numbers to decimal form:

   01010101_2 $1111_2 11001100_2$ 100000000000_2

2. How many memory cells does a 4-Kbyte memory contain?

3. How many memory cells does a 1-gigabyte memory contain? (Hint: giga is the prefix for billion, and a gigabyte must be a power of 2.)

4. Consider the addition of two decimal numbers:

   28 one's place $\rightarrow 8 + 2 = 0$ *carry* 1
   $+ 12$ ten's place $\rightarrow 2 + 1 = 3 + carry = 4$
   ―――――
   40

   Now use the same logic to compute the addition of two binary numbers as follows:

   11100_2
   $+01100_2$
   ―――――

5. Consider the architecture of the three-register machine shown in Figure 1.6. Assume that a language exists with the following instructions and their meanings:

   ```
   CP A,B -- copy the contents of register A to B
   CP B,A -- copy the contents of register B to A
   CP A,C -- copy the contents of register A to C
   CP C,A -- copy the contents of register C to A
   ADD    -- add contents of A to B; leave result
             in A
   SUB    -- subtract B from A; leave result in A
   INP    -- input a value to register A
   OUT    -- output a value from register A
   ```

A program in this language to add two numbers might be as follows:

```
INP
CP A,B
INP
ADD
OUT
```

Write a program in this language to perform the following computation and output the value of X:

$$X = 5 + 3 - 2.$$

Assume that the numbers are inputted sequentially each time the `INP` instruction is used. Do you see a relationship between the way a simple calculator works and this programming language?

6. Consider a set of wooden disks that might be used to construct a tower such as in Figure E1.1. Write a tower construction program in the spirit of the burger construction program in Section 1.2.

7. If the tower of disks is constrained to move between three posts according to a set of rules, we now have the "Towers of Hanoi" puzzle. Figure E1.2 illustrates the arrangement of the disks and

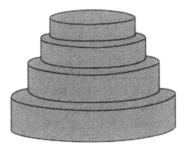

Figure E1.1

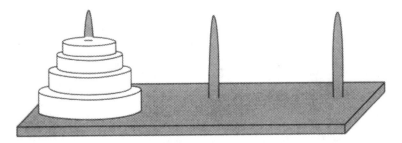

Figure E1.2

posts. The problem is to move the disks from post number one to post number three under the following conditions: 1. Only one disk may be moved at a time. 2. A larger disk may not be placed on a smaller disk. Write a tower construction program in the spirit of the burger construction program in Section 1.2. Hint: A step in the program can ask a question such as, Are all the disks moved? If the answer is no, the program can transfer to a previous step.

2 Computer Programming

omputer programming is the process by which we instruct a computer to perform a useful calculation or process. The computer can easily be described as an *idiot savant*, a term used by psychiatry to describe mental conditions in which individuals are capable of performing incredible feats of memory or calculation on request but are unable to understand the simplest activities of daily life. It is important to know that the computer will do only what it is instructed to do, no more and no less. Computers, sophisticated as they are, do not possess sentience or self-awareness. They cannot guess or anticipate what you want them to do; they simply do as instructed. There is a very old saying "garbage in, garbage out" that means if you put bad data into a program you can hardly expect to get good data out. Likewise, if your program is formed badly, the computer will not correct it for you.

Modern computers are rarely plagued by problems that cause them not to execute their instructions or to execute their instructions in a fashion other than that specified – in other words, they either work or they don't. If they seem to be acting strangely your program most likely is at fault. The most difficult task in the programming of a computer is the understanding of what the computer is supposed to do. To help make this task easier, we will examine techniques for describing problems and developing computer solutions. These techniques are useful regardless of which computer language you choose to program with.

2.1 Problem Solving and Program Development

There are many ways to approach problems. When one tries to solve problems with a computer, often the process of using the computer to solve the problem is a larger problem than that being solved! This will become clearer as you begin writing programs. Because programming is a very methodical process, one of the most efficient ways to approach problem solving with a computer is to use a well-known set of six fundamental problem-solving steps that are listed below.

---

❶ State the problem clearly.

❷ Describe resources, data needed (input), expected results (output), and the variables required for the problem.

❸ Work a sample data set by hand.

❹ Develop an algorithm to solve the problem.

❺ Code the algorithm.

❻ Test the code using the edit–compile–run cycle on a variety of data sets with known results.

Possibly the most difficult task in any kind of problem solving is the first step, that of stating the problem clearly. Not suprisingly, this step is linked to another aspect of problem solving: determining what the real problem is. An example of what we mean by this is illustrated by the following story:

> *A student and his professor are backpacking in Alaska when a grizzly bear starts to chase them from a distance. They both start running, but it's clear that eventually the bear will catch up with them. The student takes off his backpack, gets his running shoes out, and starts putting them on. His professor says, "You can't outrun the bear, even in running shoes!" The student replies, "I don't need to outrun the bear; I only have to outrun you!"*
>
> from *Strategies for Creative Problem Solving*
> by Fogler and LeBlanc, Prentice Hall, 1995.

The moral of the story is that you should know what your problem really is. In engineering, the computer is good for solving many problems, but the two most common are visualization of processes and the computation of design parameters. Of course, the computer is also good for general data management, word processing functions, and complex games. Throughout this book we will be looking at how to program and develop problem-solving strategies that utilize the power of the computer as a problem-solving aid. Because this is a book on programming, we will concentrate more on how to write a program for a given problem as opposed to trying to figure out how to solve problems in general. Nevertheless, taking the time to think over problem strategies is useful in any kind of problem-solving situation.

The problem-solving steps listed above can be applied to the solution of computing the simple statistics of a set of sample data. Specifically, we want the mean, standard deviation, and variance to be computed. The sequence of steps for this problem is as follows:

❶ State the Problem Clearly

1. Compute the mean, standard deviation, and variance for a set of sample data

This problem involves statistics, which you may or may not be familiar with. Statistics is a branch of mathematics that studies the analysis and interpretation of data. The use of statistics is an essential aspect of engineering and is important to all engineering fields. The mean of a set of data is just the average value. Engineers use the mean in many applications to predict the behavior of systems over time or the expected parameter of a component within a large group or sample set of components. For example, the number of stress fractures in a bridge or the resistance values of resistors are statistical quantities. Standard deviation and variance are measures of dispersion of the data from the mean, or how far the individual data sample values lie from the average value. Standard deviation is the square root of the variance and is related to the shape of a normal curve. The variance is the sum of the squared differences of each datum from the mean and can therefore tell us how broad, or dispersed, the data samples are from the average value.

❷ **Describe Resources, Data Needed (Input), Expected Results (Output), and the Variables Required for the Problem**

2. The formulas for mean, standard deviation, and variance can be found in any introductory statistics text and in many engineering reference and text books. The formulas are as follows:

Mean formula:

$$\mu = \frac{1}{N} \sum_{i=1}^{N} d_i$$

Standard deviation:

$$\sqrt{\sigma^2}$$

Variance:

$$\sigma^2 = \frac{1}{N} \sum_{i=1}^{N} (d_i - \mu)^2$$

For our example, we can produce statistics on the percentages of humans over 65 years old (in 1950) living in the Sunbelt. This sort of data can be found in almanacs and is given below:

AL–6.5	AK–7.8	FL–8.6	LA–6.6	NM–4.9
AZ–5.9	CA–8.5	GA–6.4	MS–7.0	TX–6.7

These data are the input to our algorithm. Each Sunbelt state (Alabama, Arkansas, Florida, Louisiana, New Mexico, Arizona, California, Georgia, Mississippi, and Texas) has a value associated with it that represents the percentage of people over 65 living in the state in 1950. The computer will only need this value rather than the information that the value represents a percentage of people or that it is associated with a state. These aspects of the data have no explicit relationship to the statistics. We can call the input variable d, the sample data, just as it is in the mean formula. Note that this variable has a subscript i that indicates a range of samples up to N. These will be specified as intermediate variables to be used by the algorithm. The outputs of the program will be the mean (the average of data), variance (the deviation from average), and standard deviation (the measure of dispersion, the root of the variance). The output variables, from the formulas, are μ, σ, and σ^2. Because neither C nor Fortran can use Greek letters, we will have to use different names for these variables.

❸ Work a Sample Data Set by Hand

3. Using a calculator, we apply the formulas using the data to get the following results:

(sum percentages)

$6.5 + 5.9 + 7.8 + 8.5 + 8.6 + 6.4 + 6.6 + 7.0 + 4.9 + 6.7 = 68.9$

(divide by number of states to get mean)

$68.9/10 = 6.89$

(use formula to compute variance)

$(1/10)((6.5 - 6.89)^2 + (5.9 - 6.89)^2 + (7.8 - 6.89)^2 + (8.5 - 6.89)^2$
$+ (8.6 - 6.89)^2 + (6.4 - 6.89)^2 + (6.6 - 6.89)^2 + (7.0 - 6.89)^2$
$+ (4.9 - 6.89)^2 + (6.7 - 6.89)^2) = 1.07$

(compute standard deviation from square root of variance)

$\sqrt{1.07} = 1.03$

✓ We perform this step in order to have results to check the computer with after we program it. You may have heard the expression: "It was lost in the translation." As with natural languages (what we use to communicate with each other), this is also true for computer programs. We will examine errors in more detail later, but it is important to remember that the computer will do whatever you tell it to do. In other words, even though your instructions may be in a form understandable to the machine, what you told it to do may not result in a correct and accurate answer! Therefore, always check the computer results against your own results.

This step is often misunderstood. It may seem incongruous with the problem-solving process to work the problem in order to solve it. The real problem is not solving for simple statistics but rather the development of a computer program to solve for the statistics. Once the program is written, *any* set of data can be inputted, and the statistics will be produced. For our example, we used ten samples. This number is easily processed using the simplest of hand calculators; however, once the program has been written we could develop statistics on data sets with thousands of elements such as the number of resistors produced by a factory in a few minutes. This would hardly be the kind of calculating that we would want to have to do by hand. What is important is that we check our algorithm against known data so that we can also check the computer output to verify that it is correct.

❹ Develop an Algorithm to Solve the Problem

4. This is best done with either a flowchart (Sec. 2.3) or pseudocode (Sec. 2.4)

Figure 2.1 is a flowchart for the statistics problem, and Figure 2.2 shows a pseudocode representation. Examine these now, but don't worry if they are not clear at this point. What is important is that the computer does not comprehend human speech or writing (i.e., natural language). A flowchart or pseudocode representation allows us to organize the set of steps, the algorithm, that we will later translate into the computer program. This translation process is called *coding*, and the program is called the *code*. The term is a carryover from when computers were programmed in binary before higher level languages were invented. Programs were indecipherable to anyone but the programmer.

Note that in the flowchart and the pseudocode representations of the algorithm we have defined **variables** using words in uppercase. Variables are named objects that hold data, or values, just like the variables that we use in algebraic equations. Each computer language has rules that restrict how variables are defined and used. Variables can contain data other than numerical information, such as text strings and special symbols, and there are also different classes of numerical data.

The variables defined in both the flowchart and the pseudocode are the same. The variable names are called **labels**, and you should choose labels that are consistent with the meaning of the data that they hold. You should also choose labels that conform to the restrictions dictated by the language that you are programming in. We will

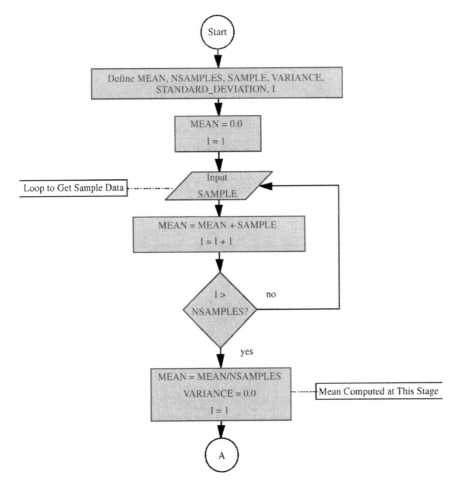

Figure 2.1a Flowchart for statistics problem.

discuss this in greater detail later, for the restrictions on labels in C and Fortran are different. You can get a feel for these differences from the program illustrations given in the discussion that follows on coding the algorithm. To summarize, a variable is a placeholder for some sort of data, and a label is the name given to the variable.

❺ Code the Algorithm

5. The codings for both the Fortran and C language are shown in Figure 2.3

Note both the similarities and the differences between the two programs. Both of these programs input the same type of data and will

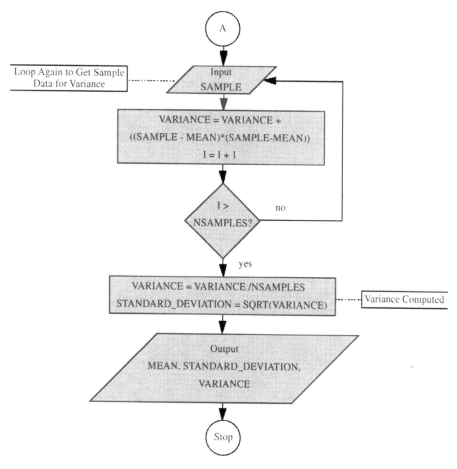

Figure 2.1b   Flowchart for statistics problem.

```
Define MEAN, NSAMPLES, SAMPLE, VARIANCE,
STANDARD_DEVIATION, I
MEAN = 0.0
loop for I=1 to NSAMPLES
   input SAMPLE
   MEAN = MEAN + SAMPLE
end loop
VARIANCE = 0.0
MEAN = MEAN / NSAMPLES
loop for I=1 to NSAMPLES
   input SAMPLE
   VARIANCE = VARIANCE +
       ((SAMPLE - MEAN)*(SAMPLE - MEAN))
end loop
VARIANCE = VARIANCE/NSAMPLES
STANDARD_DEVIATION = SQRT(VARIANCE)
output MEAN, VARIANCE, STANDARD_DEVIATION
```

Figure 2.2   Pseudocode for statistics problem.

Fortran	C

<pre>PROGRAM STATS
  INTEGER COUNT,I
  REAL DPT,MEAN,VAR,STDDEV
 	
  MEAN = 0.0
  VAR = 0.0
  COUNT = 10
 	
  DO 1 I=1,COUNT
    READ*,DPT
1 MEAN = MEAN + DPT
 	
  MEAN = MEAN/COUNT
 	
  DO 2 I=1,COUNT
    READ*,DPT
2 VAR=VAR+(MEAN-DPT)**2
 	
  VAR = VAR/COUNT
  STDDEV = SQRT(VAR)
 	
  PRINT*,MEAN,VAR,STDDEV
 	
  STOP
  END</pre>
	<pre>main(){
	  int count,i;
	  float mean,var,dp,stddev;
	
	  mean = 0.0;
	  var = 0.0;
	  count = 10;
	
	  for(i=1; i<= count;++i){
	    scanf("%f",&dpt);
	    mean = mean + dpt;
	  }
	
	  mean = mean/count;
	
	  for(i=1; i<= count;++i){
	    scanf("%f",&dpt);
	    var = var +
	    ((meandpt)*(mean-dpt));
	  }
	
	  var = var/count;
	  stddev = sqrt(var);
	
	  printf("%f %f %f\n",
	         mean,var,stddev);
	}</pre>

Figure 2.3   Fortran and C codings for the statistics program.

output identical results. Because Fortran and C compilers are available on almost all modern computers, you could generate programs on virtually any machine you wanted in either language! For now, just note the similarities and differences between the two and try to see how the flowchart and pseudocode map to both programs.

⑥ Test the Code Using the Edit–Compile–Run Cycle on a Variety of Data Sets with Known Results

Although flowcharting and pseudocoding can be performed **off-line** (without the computer), testing the code involves use of the machine itself. When you are ready to use the computer to write your own programs, you will need access to an editor, a compiler, and a linker–loader program. All of these functions may be combined in a **development environment**. Development environments are ensembles of computer software sold as a package. You should have access to such a package, or to the individual programs, before continuing with Chapter 3.

✓ When beginning to program, it is difficult to "think like a computer," but this is an important aspect of programming. Many programmers jump right in and compose their programs at the keyboard. This practice may seem expedient, but it can lead to excess errors (bugs). The optimal approach is to work out a solution first, work the algorithm in pseudocode or with a flowchart, and then go to the computer. This method helps you develop a feel for how the computer actually processes data. As you get better through practice, you will find that writing programs comes naturally to you.

2.2 The Edit–Compile–Run Cycle

The edit–compile–run cycle entails the actual development of the computer program from the algorithm. You should now understand that a computer program is a set of instructions that directs a computer to do something. How to get the computer to understand what to do is a complex process of translation from the problem that you want to solve into a language that the computer understands. The languages that we will explore in this book are C and Fortran. These languages will be discussed in greater depth later, but for now we need to explore the mechanics of getting the program into the computer and processed into a form that it understands.

From the previous section we developed a strategy for problem solving and saw what a typical Fortran and C program looks like. Examine the programs again and note that neither of them contains symbols that we would not find on a standard typewriter or computer keyboard. This is important because we have to type the program into the computer before it is able to process the program. The typing of the program is done using an **editor**, which is a special program that makes the computer work like a typewriter. The editor allows you to type words onto the computer screen instead of on paper. Editors vary in complexity and sophistication. If the editor program is capable of special formatting, italic type, boldface, and the inclusion of graphics and images, then it is called a **word processor**. The output of the editor must be what is called an American Standard Code for Information Interchange (**ASCII**) or **text-only** file. In Chapter 3 we will define ASCII in greater detail. For now, think of ASCII in terms of the symbols on a keyboard. A **file** on a computer is nothing more than a set of data stored in binary form on the **mass storage unit** of the computer. On modern computers, disk drives are used for mass storage, and the files stored on them are managed by the operating system. Recall that the operating system is a special application that manages the resources of the computer. Files on a computer, like files in an office, require special handling if data are to be managed efficiently. Each file is given a name and a specific location within the file system. Collections of files can be placed together in a folder, as if in an office, and the folder is given a name. Folders can be grouped into other folders or volumes. Volumes are like file drawers in a file cabinet. Different operating systems have different ways of representing the hierarchy of files, folders, and volumes. Some do the organization graphically to the point that files and folders have small graphic pictures on the screen (called icons) that look like paper files and folders.

The files created by the text editor contain data that represent the coding of alphanumeric characters (ASCII), and these files are processed by the compiler. Character codes that can be displayed as letters or numbers are called *printable*. When you save a program file that the editor creates, it will prompt you for a file name. The name must conform to the operating system rules for file names. You should choose names that correspond to the type of data the file contains. We call a file that contains a computer program written in a high-level language a **source code** file, or just *source*. In most systems, files with a suffix of .c contain a C program source code, and those files with a suffix of .f contain a Fortran source.

As you type your program into the computer, each character is stored as a byte of data, and the computer can interpret that byte in many different ways. In the editing and compiling stage of program development, we want the data interpreted as printable characters that represent our program. The compiler will then create an executable program that can be interpreted by the computer. This program will exist as a file called an **executable**, or executable module, and it consists of binary codes unique to the computer that the program is run on. The executable can then be run by the computer to output the data required to solve our problem. In some operating systems, a file name with the suffix **.exe** indicates an executable.

If the compiler does not understand the program that you type in with the editor, the compiler will tell you that you have a **compile-time error** and will terminate compilation. In this case it will not produce an executable, and you must edit the file to fix the error. Each programming language has a complex set of rules that must be followed when writing a program, and these rules are called the **syntax** of the language. When you learn a computer language, you are learning the syntax of that language. If you have ever learned a human language other than your native tongue, you know that you study vocabulary and grammar to speak and write the language. The vocabulary of a computer language is very small in comparison with human languages. Although it is complex, the syntax which represents the grammar of a computer language, can easily be mastered in a relatively short time (compared with the time it takes to learn a human language). Each time a program is written, the process is synonymous with the writing of a short story or novel in which the plot is the algorithm and the characters are the variables. Typically, compile-time errors are syntax errors and are easily corrected once the proper syntax of the computer language is understood.

Substantially more insidious are the **runtime errors**, or errors that occur when the computer tries to execute, or run, the compiled program. It is possible to have a program that does not violate syntax rules and is subsequently compiled into an executable but yet attempts to perform operations that are illegal. An example of such an illegal operation is division by zero. To use the short story analogy once more, it is possible to write a very poor and difficult-to-understand story without violating any of the grammar rules of a language. We will examine sources of both compile-time and runtime errors in later chapters.

The process of edit–compile–run can sometimes seem like an infinite cycle, as illustrated by the graphic in Figure 2.4. Nevertheless, the cycle is one that all must encounter while developing computer programs. The process is expedited somewhat by use of an inte-

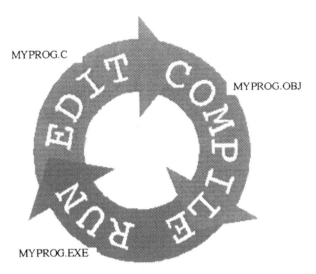

MYPROG.C

MYPROG.OBJ

MYPROG.EXE

Figure 2.4 Edit–compile–run cycle.

grated **development environment**. The development environment combines the editor and compiler functions into one program and includes the capability to instruct the operating system to run programs that have been compiled correctly. A development environment allows for rapid programming and reduces the time required to loop through the edit–compile–run cycle. Most development environments include **debugger** programs. A debugger runs a program in much the same way as an operating system does; however, the control of program execution is directly supervised by the user. Debuggers allow you to step through a program and examine or modify variables, and thus problems or errors can be detected dynamically.

2.3 Flowcharts

Flowcharting is a simple graphical method of representing program sequences and algorithms using a standard set of symbols to represent program flow. The International Organization for Standardization (ISO) compiled a set of symbols that would be standardized for computer flowcharting as the *Recommendation R1028 Flowchart Symbols for Information Processing* in 1970. This symbology conformed to the American National Standards Institute (ANSI) flowchart symbols. A subset of these symbols and their meanings are reproduced below.

ISO/ANSI FLOWCHART SYMBOLS

Process: A defined operation, or set of operations, causing a change in value, form, or location of information.

comment(s)

Descriptive clarification or comment. Dashed line extends to symbols as appropriate.

Input/output: A general I/O function.

Connector: Exit to or entry from another part of the chart.

Arrowheads and Flowlines: Flowlines link symbols and illustrate data flow. Add arrowheads to paths when linkage is not left-to-right or top-to-bottom.

Decision: Allows for program flow control. When a path enters the decision symbol, a question is asked about the state of a variable, and program flow proceeds on the basis of the response.

Predefined Process: Used for subroutine or function call. Process is defined on another flowchart.

Document: Output data to printer.

ISO/ANSI FLOWCHART SYMBOLS (cont.)

Off-Line Storage: Results of I/O are stored to disk, tape, etc.

Display: Output data to monitor.

Keyboard: Input data from keyboard.

Flowcharting is useful because it shows the behavior of the algorithm in a visual way. Engineers tend to prefer flowcharting as an algorithm representation because it mirrors the flow diagrams often used to describe physical systems such as those found in controls, fluid dynamics, communications, and manufacturing. To get an idea of how to construct a flowchart, consider an algorithm to decide whether to round a number, stored in the variable X, up or down. For simplicity, we will assume that any variable has only one significant digit after the decimal place and that, if a division of the variable is made, the least significant digits will be truncated or cut off. In other words, if X is 13.0, then X divided by 8 will be 1.6, not 1.625.

To do the rounding, if the fractional part is greater than or equal to 5, we will want to round X up to the next whole number. If the fractional part is less than 5, we will want to round down. If X is 3.2, then we want 3.0. If X is 5.8, we want 6. How can we do this arithmetically? Well, if we divide X by 10 and then multiply by ten, we will get the whole part of X. Say X is 3.2. Divide by 10. The variable X is now 0.3 (recall the truncation that takes place). Now multiply X by 10; X is now 3.0. The flowchart for this is as follows:

```
X = 3.2
F = X/10
X = X * 10
```

Note that the flowchart is just a process box. Also note that we do three things to the X variable. We set the value of X equal to 3.2. This is called *assignment* because we are assigning a value to the variable. The assignment $X = 3.2$ is called a **statement**. We then set the value of X equal to the current value (3.2) divided by 10. This computation yields the value 0.32, but we said that a variable would only hold one significant digit; thus the 2 is lost, and the value of X at this point is 0.3. We then multiply the current value of X by 10 and make that result the new value of X. This would be fine, but we have lost the original value of X. We need to determine what the fractional part of X was in order to determine whether to round up or down. For this, we will introduce a variable F to hold the fractional part of X. Consider the following process box:

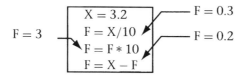

When the four assignments are complete, the value of X will be 3.2, and the value of F will be 0.2. Convince yourself that this is what is processed (remember what happens to all digits beyond the first after the decimal place).

To do the rounding, we have to ask the question, Is the value of F greater than or equal to 0.5? If it is, then we need to round X up by one; if not, we need to subtract F. To ask a question in a flowchart, a decision diamond is used. For this problem, it might look like the following:

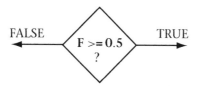

What the box does is ask the question, Is F greater than or equal to 0.5? If the answer is yes, we say that the statement $F >= 0.5$ is true. If the answer is no, we say that the statement is false. If true, then we need to round X up. This can be accomplished by subtracting F from X and then adding one. If the statement is false, then we just need to subtract F from X.

The full flowchart for the problem is given in Figure 2.5. Do you notice any redundancy in the algorithm? The flowchart makes it easy to see where program efficiencies can be gained. For example,

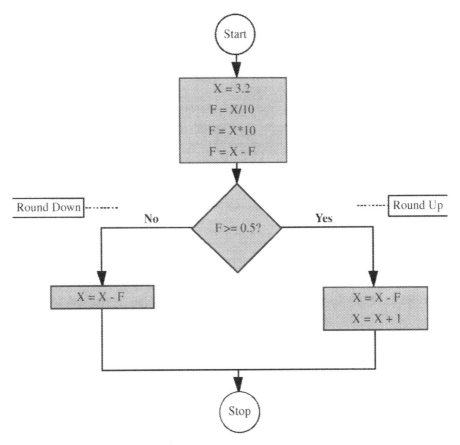

Figure 2.5 Flowchart for rounding algorithm.

we could combine the last two assignments in the first process block to be $F = X - F*10$. We cannot combine all three assignments of F. Do you see why? The first assignment presupposes that a truncation will take place that will eliminate all but the first significant digit after the decimal. A further efficiency can be gained by subtracting F from X prior to the decision diamond, for that operation will take place regardless of the decision outcome. If the subtraction is moved, then the false side of the decision will do nothing.

> ✓ **Flowcharts graphically depict the flow of information. Therefore, try to develop a visual representation of information flow through your algorithm.**

2.4   Pseudocode

Pseudocode, or false code, is a simple, structured representation of a program sequence or algorithm that is not intended to be run on a machine. The key to good pseudocode is the use of clear, easily understood phrasing. Pseudocode is, loosely, a flowchart without the graphics.

There are no strict rules to pseudocode, but the simpler the statements the better. A list of basic pseudocode terms is as follows:

DEFINE	used to specify variables
LOOP	starts a looping or iterating activity
END LOOP	marks end of the loop
IF, THEN, ELSE	defines a conditional
INPUT	input of data
OUTPUT	output of data

Whether to use pseudocode or flowcharting is an individual preference. The advantage of flowcharting is that you can easily visualize the flow of the program, whereas the downside is the tediousness of construction. Pseudocode has the distinct advantages of being easier to write down and faster to convert into the computer program. The disadvantage is that pseudocode is more difficult to analyze when problems arise in the algorithm. For some algorithms the pseudocode is very complex. You should practice and use both methods while you are learning programming skills.

The pseudocode for the rounding example might be the following:

Start
    $X = 3.2$
    $F = X/10$
    $F = X - X*10$
    $X = X - F$
    if F is greater than or equal to 0.5 then $X = X + 1$
Stop

Note that the efficiencies discussed earlier have been added. Also, the if statement could have been written more compactly as *if F* $>=$ 0.5 *then* $X = X + 1$.

2.5 Program Structure

An analogy presented earlier said that a computer program is like a story or essay. Like good writing, a computer program must have structure. There must be a beginning, or prologue, a body, and an ending. Like any good piece of writing, these pieces should fit together well. Examine the structure of a C program to output the words *Hello World!*, as shown in Figure 2.6. The bold highlighted items are elements of the program that must be present; thus, you can see that the simplest C program will be just **main(){ }**. This program will be compiled and run, but it will not do anything because no program statements have been made. The statement is an instruction to the computer to do something, and it is perfectly fine for the computer to do nothing, although it is not very logical. Conversely, if any part of **main(){ }** is missing from a program, it will not be compiled, because this expression represents the fundamental structure of the C program.

The first line of the program in Figure 2.6, **#include <stdio.h>**, is what is known as a **compiler directive** and is optional in many programs. A program called the **C Preprocessor** evaluates the compiler directives before actual compilation. Compiler directives begin with a pound-sign character (#) in the first column. The **include** directive instructs the compiler to add the contents of a file to the source being compiled (in this case, the file "stdlib.h"). This file contains information about the standard library that must be included to perform I/O operations. Consult Appendix D for more in-depth discussion of the C preprocessor and a sampling of the various compiler directives that it recognizes.

The second line of the program is a **comment**, which is a note that the programmer puts in the program to describe what the program is doing. The C comments begin with /* and end with */. Any characters

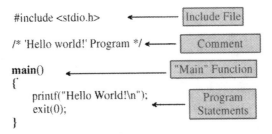

```
#include <stdio.h>              ◄————— Include File

/* 'Hello world!' Program */ ◄————— Comment

main()              ◄————— "Main" Function
{
    printf("Hello World!\n");
    exit(0);                    ◄————— Program
}                                         Statements
```

Figure 2.6 C program structure.

between these symbols are ignored by the compiler. This feature allows the programmer to add annotations to help in the understanding of the program.

> ✓ How do I know when I have commented my program adequately? Many beginning programmers ask this question, and the answer is not straightforward. However, two rules of thumb can be followed that will aid you in commenting your programs:
>
>   [1] Comment so that another person with your level of programming expertise can understand what the program is doing.
>
>   [2] Comment so that when you read the program you will be able to understand what you did after a considerable period has passed!

The body of the program contains statements (*printf* and *exit*) that instruct the computer to send the characters *Hello World!* to the console and then exit. The statements, or computer instructions, are constrained by the syntax of the language being used. We explore the various kinds of statements in the chapters following this one; however, it is important to note that a C statement is always followed by a semicolon (;). The semicolon tells the compiler where the statement ends; omitting the semicolon will generate compile-time errors.

The structure of a Fortran *Hello World!* program is shown in Figure 2.7. From the bold highlighted words we can see that the simplest Fortran program is just **PROGRAM END**. In all computer languages, a program without statements is called a **program shell**. You can start with a shell and just add statements until you have a program. In the case of a C program, you must always be careful to have the two braces { }, or the compiler will generate errors.

Fortran has an additional restriction that does not apply to C: the requirement that lines of Fortran code conform to a specific column placement constraint. This constraint is shown in Figure 2.8, which illustrates the column fields of a Fortran statement. A 'C' in column 1 of a Fortran statement indicates a comment, and any subsequent characters are ignored. If the statement is not a comment, columns 1 through 5 are reserved for the statement number. Not all statements require a number. Any character in column 6 indicates a continuation of the previous line (i.e., statements requiring more

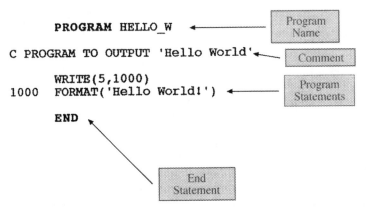

Figure 2.7 Fortran program structure.

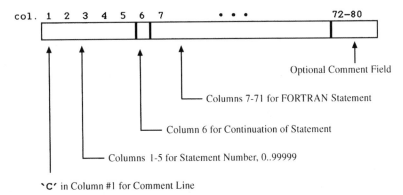

Figure 2.8 Fortran column restrictions.

space than that allowed in columns 7 to 71). Columns 72 to 80 are reserved for an optional comment field. The Fortran column restrictions are a carryover from the days when Fortran program statements were punched onto paper cards. Modern Fortran compilers (Fortran-90, for example) no longer have the column restrictions and use the "free field" format, like C.

You should now try to **edit–compile–run** what is called the C "Hello World!" program. Type the program into your editor exactly as it is printed in Figure 2.6, being particularly careful to type a backslash-n (\n) in the printf statement. Save the source file as **hello.c**. Your compiler should produce an executable with a different filename, and if you typed the program exactly as specified, it is unlikely that you will encounter a compile-time error. Compilers typically will generate an executable file named **hello.exe** or just **hello**. To run your program, type the program filename or click on the program icon (this will vary depending on the operating system you are using). If you are using a development environment, you should be able to run the program from within it.

You may want to experiment with the compiler to see what errors look like. Leave off the right-brace (}) and try to compile. Try misspelling or capitalizing **main**. Try leaving off the */ at the end of the comment. In Chapter 3 we will see how data are specified, variables are created, and arithmetic and logical statements are formed.

REVIEW WORDS

ASCII
C preprocessor
comment
compile-time error
compiler directive
debugger
development environment
edit–compile–run
editor
executable
file
label
mass storage unit
off-line
program shell
runtime error
source code
statement
syntax error
text-only
variable

EXERCISES

1. List the first four problem-solving steps to solve the following:
   A fluid pump curve can be approximated by the following

equation:

$$h_A = -1.22 \times 10^{-5}Q^3 + 164(-0.16Q^3 + 50)$$

Over the range 0–175 ft^3/min, where Q is in ft^3/min and h_A is in ft, a fluid system curve is given by

$$h_f = 3.52Q^2$$

How can the operating point be determined?

2. Draw a flowchart for Exercise 1.

3. Write a pseudocode program for Exercise 1.

4. Ohm's law is given by the following equation:

$$E = IR$$

where E is the voltage in volts, I is the current in amperes, and R is the resistance in ohms. Draw a flowchart for an "Ohm's law calculator" that allows the user to input any two parameters and to output the third.

5. Write a pseudocode program for Exercise 4.

6. What compile-time error is generated when the following program is compiled?

```
#include<stdio.h>
main()
{
     printf("Hello World!");
     exit(0);
}
```

What is missing from the program?

7. What type of error occurs when the following is compiled?

```
#include<stdio.h>
main()
{
     pri("Hello World!");
}
```

The error is neither compile-time or runtime.

3

Types, Operators, and Expressions

Common pitfalls in engineering programming are the failure to observe type restrictions and the lack of understanding of operators and expressions with regard to type. This chapter explores types, operators, and expressions in both Fortran and C and seeks to compare the two languages and explain how they each function in this domain. The contrast will foster a deep understanding of this most critical aspect of programming.

Computer languages are defined in terms of their syntax and their **reserved word** list. A reserved word is one that the compiler recognizes as a unique descriptor of the language, and thus the word may not be used out of the context of a program statement. Fortran uses many reserved words. Because our interest is in reading Fortran and not programming it, we will simply identify the words as they appear in programs. In contrast, C has a very short list of reserved words, and they should be committed to memory. The C reserved words are (loosely grouped in order of appearance in the text) as follows:

C Reserved Words

int	if	return	struct
float	else	void	
double	for	extern	union
char	do	auto	typedef
long	while	static	
short	break	register	
unsigned	continue		
	switch		
sizeof	case		
	default		
	goto		

In C and Fortran, variable names may consist of any combination of letters and numbers, but they must begin with an alphabetic character. In C, the underscore (_) is considered to be an alphabetic character. The length of the name is compiler-dependent, but names that exceed 32 characters should not be used, for these will complicate the reading and understanding of the program. When discussing Fortran, we will follow the early conventions of only allowing uppercase letters. In C, variable names are **case sensitive**, meaning that a variable named SLEW_RATE is distinct from the variable named Slew_Rate. A set of variable names is presented here to clarify this concept. The names in the left column are valid in C, the names in the right column are similar to those in the left but are not valid names and would produce a compile-time error if used. Note that what makes a name valid or not can be just a subtle difference, but invariably it is the use of a number as the first character in the name or the use of a forbidden character (* and ˆ in the listing).

Valid C Variable Names	Not Valid
X1	1X
x95a	95x
slew_rate	slew*rate
PID_control_gain	PIDˆcontrolˆgain

In mathematics, we use variables and operators to form expressions. In programming, like mathematics, an **operator** defines an operation on a variable or between two variables. An **expression** is a combination of variables and operators that forms a result based on the definitions of the operators and the types of the variables. For example, the expression $5 + 6$ results in the value 11. The expression is the addition of the integers 5 and 6. We call the computer processing of an expression **evaluation**. How the expression is evaluated depends on specific rules of syntax and order of operations. We begin our study of expression evaluation by first defining the different types of variables possible in C and Fortran programs.

3.1   Data Types

Type is a form of classification or grouping of variables. We assign types to data classes, such as integers or real numbers, and there are specific rules that govern how the types are utilized in the program. These rules apply when forming program expressions and statements and define the results that one obtains when types are mixed. Consider the following three numbers:

5   5.3   0

Mathematically, we can say that all three of these numbers are decimals and exist within the span of real numbers. The number 5, however, is also a member of the integers, as is the zero, whereas 5.3 is a member of the set of real numbers and is not an integer. In mathematics, the context of how the numbers are used indicates to us how we should interpret them. The computer is no different; however, the span of interpretation is limited and specific. Furthermore, the interpretation depends on the type of the variables being used and the nature of the expressions that the variables and numbers appear in.

Consider the following mathematical expression:

$$X = 5$$

We state here that X is equal to 5, and thus when X appears we may also consider it to be 5. In computer terminology, we say that X has been *assigned* the value 5. The equal sign is called the **assignment operator**, and this should be distinguished from use of the equal sign as an equivalence operator as encountered in algebra (meaning that something is *equivalent* to something else). To understand this, consider the following expression:

$$X = X + 5$$

There are two interpretations: that $5 = 0$, an illegal conclusion based on the rules of arithmetic, or that X is equal to the previous value of X added to 5 to become a new value of X. The computer follows the second interpretation because it *assigns* the value of $X + 5$ to X, effectively creating a new value for X without solving the algebraic expression for a value for X. Section 3.4 discusses assignment operators in more detail. At that time we will introduce the idea of the right- and left-hand sides of an assignment expression. For now, it is enough to understand that the evaluation of the expression on the right of the assignment operator *replaces* the value of the variable to the left of the operator.

Something of a dilemma arises when one encounters the following sequence:

$$X = 3$$
$$X = X + 5.3$$

Of course, we immediately say, "There is no dilemma, X has been assigned the value $3 + 5.3$, or 8.3!" This conclusion would be based on our previous discussion in which we saw that the equal sign meant assignment and not algebraic equivalence. The conclusion is only partially correct. The problem arises with the computer interpretation of the two expressions (recall that an expression is a combination of variables and operators). In the C and Fortran computer languages, one must ask what the **type** of X is. If X is of real (i.e., floating point) type, then there is no problem, and X is now assigned the value of 8.3. However, if X is of integer type, then the newly assigned value will be 8, and information may be lost. The computer cannot resolve the rational portion of the 5.3 if it is to keep X as an integer.

When we want to use a **variable** in a program, we must assign a label to identify it, and, when programming in C, we must explicitly *declare* what type the computer should assign to the variable. There are four fundamental data types in the C language and six in Fortran. The C data types are given in the table below:

Type	Description	Example
int	integer; counting numbers	$\ldots -2, -1, 0, 1, 2 \ldots$
float	real; floating point numbers	$3.1416; -0.0003$
double	double-sized reals	larger values
char	single-byte characters	A, b, &, $*$

Fortran uses similar typing where INTEGER $\equiv^*$ int, REAL $\equiv$ float, DOUBLE PRECISION $\equiv$ double and CHARACTER $\equiv$ char. Fortran adds the additional types COMPLEX and LOGICAL. The Fortran LOGICAL is used to hold a Boolean TRUE or FALSE value. The C equivalent to a Fortran LOGICAL variable is any variable because C interprets a Boolean false as a zero value and any other value as true (more on this in Section 3.3). The Fortran COMPLEX data type has no C

---

*The symbol $\equiv$ means "is equivalent to."

```
main(){
    int i, j, k;
    float x_pos, y_pos;
    char c;
        ⋮
}
```

Program 3.1 Declarations.

equivalent; however, in Chapter 6 we will see how complex numbers can be represented as data structures in C.

The declarations of variables must appear at the beginning of the program. Consider C Program 3.1, which has a declaration for integer variables i, j, and k; floating point variables x_pos and y_pos; and a character variable c. If we try to use variables in the program that are not declared, the compiler will not recognize them and a compile-time error will result.

There are three type modifiers in C for integers (**long, short**, and **unsigned**). A **long int** is used to declare an integer variable with twice the numerical capacity of an **int** (see the earlier discussion of **float** and **double**). The **short int** declares an integer variable with half the capacity of an **int**. An **unsigned int** is a positive only variable and is often used to represent binary numbers. The numerical capacity of the variables is dependent on the compiler and the host machine architecture.

Typically, one encounters the following variable sizes on computer workstations:

Type	Size in bytes	Numbers represented
short int	2 bytes	$2^{16} = 65,536$
int	4 bytes	$2^{32} = 4,294,967,296$
long int	8 bytes	$2^{64} = 1.8 \times 10^{19}$

Note that a signed integer, the default, must use up half of its number of values represented for the negative numbers and half for the positive (zero is considered a positive number in this scheme). A two-byte **short int** will represent numbers in the range $-32,768 \rightarrow 32,767$. An unsigned *short int* will represent $0 \rightarrow 65,535$. The internal representation used for signed integers is called two's complement, and any further discussion of it is beyond the scope of this text. Nevertheless, to determine the numerical range of an integer variable you must know the size in bits of the variable and whether or not the variable is signed or unsigned. If you know the number of bytes in the variable, then this number can be multiplied by 8 to determine the number of bits. The total number of values that the variable can represent is determined by raising 2 to the power of the number of bits. Thus, if a variable is 2 bytes, it is 16 bits in size. That variable can represent 2^{16} values, or 65,536 values. If the variable is unsigned, it can have values in the range of $0 \rightarrow 65,535$ like the *short int* described above. If the variable is signed, the range is $-32,768 \rightarrow 32,767$ (the lower, negative value is one-half the total number of representations, and the upper, positive value is one less to account for the zero). In Section 3.5 we will introduce the C *sizeof( )* operator and show how the number of bytes of a variable can be dynamically determined by the computer from within a C program.

Floating point numbers are represented very differently than integers, but the discussion of how this is done is beyond the scope of this text. Nevertheless, it is possible to determine the range of values, for both **float** and **double** variables from the "float.h" and "limits.h" include files that are supplied with the C compiler. This topic is explored further in Chapter 5.

A variable declared as **char** stores a single byte of data. A **char** variable stores an integer from -128 to 127, whereas an **unsigned char** stores from 0 to 255. Character variables can be treated like small integers in arithmetic or logical expressions, or they can store a single **ASCII** character, which is a coding scheme for representing alphanumeric, printing, and control characters with a byte of data. For example, the character A has the following equivalences:

$$A \equiv 65_{10} \equiv 41_{16} \equiv 101_8 \equiv 01000001_2$$

The computer sees an A as the binary number 01000001_2. We can use either decimal, hexadecimal (base 16), octal (base 8), or a character constant when working with character variables. Note that a lowercase a is different from the uppercase:

$$a \equiv 97_{10} \equiv 61_{16} \equiv 141_8 \equiv 01100001_2$$

Fortran uses the reserved word CHARACTER along with a size value to specify a character variable. For example, the following declaration would form a Fortran character variable of ten characters called *NAME*:

```
CHARACTER*10 NAME
```

The 10 following the asterisk specifies the number of characters in the variable. A C character variable contains a single character value, whereas a Fortran character variable can represent any number of characters up to a preset limit of the compiler (typically 32,767). For C to represent more than one character in a single variable it is necessary to form a subscripted character variable. The subscripted character variable is called a string, and we will explore it further in Chapter 6. The Fortran equivalent of a nonsubscripted C character variable is shown below for the variable name *single*:

```
Fortran declaration     C declaration
 CHARACTER*1 SINGLE      char single;
```

We can represent constants in a variety of ways in C, and these are illustrated in Table 3.1. Variables may be initialized to constant values at declaration time, and Program 3.2 shows the initializations of variables that were introduced in Program 3.1. Note that the variable *k* has been left unspecified at program start. We must assume that it will be given a value later in the program. If it is not given a value at some point in the program, the compiler will issue a warning that the variable was never initialized. This occurs often during program development when variables may be declared but not used until later programming is accomplished. When this warning appears after program completion, it may indicate that an extra variable is

Table 3.1   *C Constants.*

Integers	10	−10	65536	0
Floats	10.2	3.1416	−10.3295	0.0
Scientific notation[a]	10.2e10	−10.2e−4	3.18776e23	
Character[b]	'A'	'a'	'$'	
Hexadecimal[c]	0x41	0x61	0xABCD	
Octal[d]	0101	0141	01234567	

[a]e or E used to indicate exponent.
[b]Single quote ' used to identify single-character constant.
[c]0x precedes a hexadecimal constant.
[d]0 precedes an octal constant.

```
main(){
   int i=0, j=-10, k;
   float x_pos= 0.0002, y_pos = 3.24e-2;
   char c=`T';
        .
        .
        .
}
```

Program 3.2 Initialization of variables.

present that was never used. This may or may not have consequences for the algorithm, and the unused variable should be removed if it has no role in the solution of the problem.

The value of the character variable *c* was initialized to the character constant *T*. The internal representation of this variable will store the number 84, which is the ASCII code for the letter T (consult the ASCII tables in Appendix C). Character variables may be treated as small, one-byte, integer variables. In this fashion, arithmetic operations can be performed on sequences of letters, as we will see in later discussions.

Because the use of constants in Fortran will be obvious from the context of the programs encountered, specific initializations are not shown.

3.2   Arithmetic Operators

There are three primary classes of operators in the C and Fortran languages: arithmetic, logical, and relational. Arithmetic operators work with numerical quantities and produce numerical results. The arithmetic operators for each language are shown in Table 3.2. From the table you should note that C does not have an exponentiation operator and that Fortran does not have a modulus operator. Fortran was developed to code mathematical expressions; hence, the power (exponentiation) operator is included. The C language was developed to write computer operating systems in which the modulus is used heavily in sorting and searching tasks. This does not mean that Fortran has no modulus or C no exponentiation. Both languages include functions for these tasks. In Fortran, A modulo B is computed by the MOD(A,B) function, and in C the pow(x,y) function computes the power of x to the y. The use of functions is explored in greater detail in Chapter 5.

Table 3.2 *Arithmetic Operators.*

	C		Fortran
Addition	+	Addition	+
Subtraction	−	Subtraction	−
Multiplication	*	Multiplication	*
Division	/	Division	/
Modulus	%	Exponentiation	**

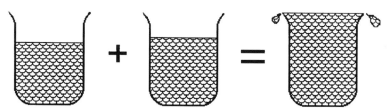

Figure 3.1 Overflow.

Because we are working with computers of finite memory, arithmetic operations are subject to **overflow** and **underflow**. Basically the terms are the same: overflow refers to positive results, and underflow to negative. Consider Figure 3.1. Two beakers, each over half full, are added together in a third: the result is overflow. Now assume we have three 1-byte unsigned integer variables a, b, and c (we could do this in a C program by declaring the variables as type **char**). We know that these variables can only hold the values 0 to 255. If we make $a = 128$ and $b = 128$, what happens when we compute $c = a + b$? The computer responds with $c = 0$! This may be easier to see from the following diagram:

	bit 8	bit 7	bit 6	bit 5	bit 4	bit 3	bit 2	bit 1	bit 0
$a = 128$		1	0	0	0	0	0	0	0
$b = 128$		1	0	0	0	0	0	0	0
$c = 256$	1	0	0	0	0	0	0	0	0

The first column shows the values of the variables in decimal and the equivalent values in binary, which is how the computer stores the numbers. Note that c is indeed equal to 256 (2^8), but our variable "runs out of bits" and overflows because only 8 bits, bit 0 through bit 7, are available to the variable. The two bits in the bit 7 position add, but a binary add of $1 + 1$ is not 2: it is zero with a carry of 1. Hence, the carry is lost in the computation, and the result left in the variable c is zero.

What happens if $a = 128$ and $b = 130$? This result is as follows:

	bit 8	bit 7	bit 6	bit 5	bit 4	bit 3	bit 2	bit 1	bit 0
$a = 128$		1	0	0	0	0	0	0	0
$b = 130$		1	0	0	0	0	0	1	0
$c = 256$	1	0	0	0	0	0	0	1	0

Once again the carry is lost, and now the variable c has taken a value of 2, which is hardly the arithmetically correct value of 258. This exercise should underscore the importance of our problem-solving Step 3 (to work a sample data set by hand). **You must be aware of the magnitudes of your computed results.** Integer overflow is somewhat predictable, but the response of the computer to floating point overflow will vary, depending on the machine. Some machines give unpredictable results, whereas others provide an error message at runtime.

Underflow results when a subtraction exceeds variable limits. When subtracting, the computer negates the subtracted number and then adds. In other words, $C = A - B$ becomes $C = A + (-B)$. Recall that signed numbers in the computer reduce the range of possible values represented. As with overflow, strange results can occur when subtracting if underflow occurs.

> ✓ When in doubt regarding variable sizes, use *double* for floating point and *long int* for integers. This guarantees that the results will compute to the maximum possible binary variable size.

Although multiplication and division may also suffer from overflow and underflow, a more prevalent problem with division is **divide-by-zero**. A divide-by-zero, when it occurs, will result in a runtime

Table 3.3 *Comparison of*
Modulus to Division Operator.

i	j	i % j	i/j
1	4	1	.25
2	4	2	.5
3	4	3	.75
4	4	0	1
5	4	1	1.25

error. Some compilers will warn when a divide-by-zero is suspected, but in most cases the problem will appear when a value computes to zero. Program 3.3 will generate a divide-by-zero error when the expression $m = 1/(i - 2 - j)$ is evaluated. From this example, you may wonder why the compiler does not catch the divide-by-zero, for it is obvious to anyone looking at the program who understands simple arithmetic. It is important to realize that the compiler does not evaluate the statements; the computer does this at runtime. The compiler is only concerned with translating the program from the high-level language into a coding that the computer can understand.

Integer division results in a **truncation**, or chopping off, of any fractional part of a division. The value of $x = 5/2$ is $x = 2$. This is a common error that is discussed further below. The modulus operator is only defined for integer variables and computes the remainder of an integer division, where $i\%j = \text{remainder}(i/j)$. Table 3.3 shows the difference between a modulus operation on two numbers and a division. The modulus operator can be used to produce correct results from integer division without conversion to floating point.

Precedence rules determine the sequence of how operations are evaluated by the compiler. These rules are discussed in Section 3.9 and are listed for C and Fortran in Appendix E. You can, however, force evaluation by the use of parentheses, in both languages. Any expressions in parentheses are evaluated first, and when parenthesized expressions are nested, the deepest-nested pair is evaluated first; then the second deepest, and so on. Without knowing the precedence rules, one might find the following expression ambiguous:

$$\text{num} = 5 * 3 * 2 + 1 * 10$$

```
main(){
    int i, j, m;

    i = 4;
    j = 2;
    m = 1/(i-2-j);
        ⋮
}
```

Program 3.3 Divide-by-zero error.

The compiler will evaluate this to *num* = 40 because multiplication precedes addition and associates left to right. The desired value may have been *num* = 450, which is computed from the following:

$$num = (5 * 3) * (2 + 1) * 10$$

✓ **When in doubt regarding precedence, force the evaluation (and add clarity to your expression) with parentheses.**

Of more immediate importance is the **promotion** of type that can occur in arithmetic operations of mixed type. When an operation is specified between two types, the lower type (determined from the hierarchy below) is *promoted* to the higher type before the operation is computed.

```
float          → double
long           → double
int            → long
char, short    → int
signed         → unsigned
```

You must be very careful when combining integer and floating point variables. If an integer is made equal to a float, the fractional part will be truncated. The most common error is generated by statements such as the following, where *x* is a float variable:

$$x = 5/2;$$

The computer will assign x the value of 2.0. This happens because 5 is an integer and so is 2. When they divide, they divide as integers, and the remainder is lost. The integer result, 2, is then promoted to float because x is a float variable and becomes 2.0! Any of the following will assign x an arithmetically correct value of 2.5:

$$x = 5.0/2; \quad x = 5/2.0 \quad x = 5.0/2.0;$$

In the first case, the presence of the float constant 5.0 *promotes* the integer 2 to float before the division (the reverse occurs in the second case). The third case is the preferred representation in which both constants are specified as float. In Chapter 5 we will examine type conversion in greater detail and see how conversions can be controlled more explicitly.

3.3 Logical and Relational Operators

Logical and relational expressions formed with their respective operators evaluate to what is called a **Boolean result**, or true or false. In C, *zero* is a Boolean false, and any nonzero value is considered to be a Boolean true. In Fortran, a Boolean false is defined as the Boolean constant .FALSE., and a Boolean true is defined as the Boolean constant .TRUE. Logical operations are defined by Boolean logic, as illustrated in Table 3.4. Exhaustive discussion of Boolean logic is beyond the scope of this text; however, the basic operations of NOT, AND, and OR are easy to commit to memory. The Boolean NOT simply changes true to false or false to true. For an AND operation to be true, all values must be true; otherwise, the value is false. With OR, all values must be false for the operation to be false; otherwise, the value is true.

Table 3.4 *Boolean Logic.*

A	B	not A	A and B	A or B
false	false	true	false	false
false	true	true	false	true
true	false	false	false	true
true	true	false	true	true

Table 3.5 *C and Fortran Logical and Relational Operators.*

C		Fortran	
Negation	!	Negation	.NOT.
And	&&	And	.AND.
Or	‖	Or	.OR.
Equivalence	==	Equivalence	.EQ.
Not Equal	!=	Not equal	.NEQ.
Greater than	>	Greater than	.GT.
Less than	<	Less than	.LT.
Greater or equal	>=	Greater or equal	.GE.
Less of equal	<=	Less or equal	.LE.
True	nonzero	True	.TRUE.
False	0	False	.FALSE.

Table 3.6 *C Logical and Relational Expression Evaluation* ($a = 10$ $b = 5$ $c = 0$ $d = 5$).

$!a$	evaluates to →	0	.NOT.a
$!c$	evaluates to →	1	.NOT.c
$a \&\& b$	evaluates to →	1	a.AND.b
$a > b$	evaluates to →	1	a.GT.b
$b > a$	evaluates to →	0	b.GT.a
$a == d$	evaluates to →	0	a.EQ.d
$a = d$	evaluates to →	5	a = d
$d >= b \&\& c < a$	evaluates to →	1	d.GE.b.AND.c.LT.a
$a > b ‖ c > b$	evaluates to →	1	a.GT.b.OR.c.GT.b

The logical and relational operators for each language are listed in Table 3.5, and the expressions formed from these operators are used in the control of flow in the program discussed in Chapter 6. To summarize this section and prepare for what is to come, we present a set of examples of logical and relational expressions and their subsequent evaluation by the computer in Table 3.6. In this table we show relationships between four integer variables with fixed values. These variables with their values are $a = 10$, $b = 5$, $c = 0$, and $d = 5$. The first column of the table has various expressions of one or more of the variables in C. The third column indicates how the computer would evaluate the expression. The last column shows the equivalent expressions in Fortran. Note that the result of Fortran evaluation of Boolean expressions is a Boolean constant (.TRUE./.FALSE.), not zero or one. In Fortran, the Boolean result of an expression can only be used in a program control statement or as an assignment to a Boolean variable.

Note the evaluation of the expression $a = d$ in the seventh row. This illustrates a common error when forming conditional expressions in C (see *if statements*, Chapter 4). Because the evaluation is nonzero, the computer will interpret it as true. If you had intended a test to see if a was equal to d, this expression would evaluate incorrectly. The expression is an assignment and not a relational test. If the programmer had wanted to determine if a was equivalent to d, then the expression $a == d$ should have been used.

✓ Observe caution when using the equivalence (==) operator. It is easy to miss typing both equal signs, and the compiler will not signal an error.

3.4 Assignment Operators

In the first section of this chapter we discussed the assignment operator, the equal (=) sign, as indicating the replacement of a variable with a new value. If we wish to pose the question, Are two expressions equal?, we use the equivalence (==) operator in C or the equals (.EQ.) operator in Fortran, as discussed in the previous section. It is necessary at this point to introduce the C concept of **left-hand side** (lhs) and **right-hand side** (rhs), which refer to the sides of an assignment expression, as shown in Figure 3.2. The lhs of an assignment must be an addressable quantity, such as a declared variable. For example, a common mistake that the first-time programmer often makes is a statement such as the following:

$$X + Y = Z + 3.8;$$

This is perfectly acceptable in algebra, but it represents an ambiguity to the computer. This statement is telling the computer to

Table 3.7 *C Examples of C op=*
Assignments.

$x = x + 5$	$\rightarrow$	$x \mathrel{+}= 5$
$x = x * (y + 3)$	$\rightarrow$	$x \mathrel{*}= (y + 3)$
$x = x/q$	$\rightarrow$	$x \mathrel{/}= q$
$x = x \% 16$	$\rightarrow$	$x \mathrel{\%}= 16$

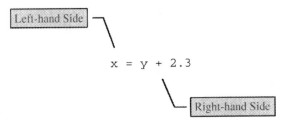

Figure 3.2 Sides of an assignment expression.

replace the value of $X + Y$ with $Z + 3.8$, an action that makes no sense. A common compile-time error is "LVALUE required," which indicates that something is wrong with the lhs of the expression in question. Likewise, constants may not appear on the lhs. The following will generate an error because the computer is being asked to redefine the value of a constant, which is a fixed quantity:

$$5 = X + Y;$$

The C language has an additional form for the assignment operator called the $op=$ (op equal), where op can be any of the arithmetic operators ($+ - */\%$). The $op=$ works for assignment expressions of the form

$$X = X \; op \; \text{<expression>}$$

which become

$$X \; op= \; \text{<expression>}.$$

Table 3.7 shows examples of each of the forms of the $op=$. Fortran has no equivalent to this operator. The use of the $op=$ operator is optional, but in many cases it can make the program more readable and compact.

3.5 Unary Operators

Unary operators work with a single variable or constant. The most well-known is arithmetic negation, the minus sign ($-$). In some cases the unary plus sign (+) may be used for program clarity, but positive is the assumed sign of constants. Both of these operators are used in Fortran and C; however, C provides three additional unary operators that Fortran does not. The C increment (++) and decrement ($--$) operators are used either to add one or subtract one from a variable. These operators provide the following equivalences:

$$X = X + 1; \Leftrightarrow X+ = 1; \Leftrightarrow ++X;$$
$$Y = Y - 1; \Leftrightarrow Y- = 1; \Leftrightarrow --Y;$$

Use of the increment–decrement operators allows for a very compact representation when incrementing–decrementing a variable. The operators may also be used **postfix**, or after the variable, as follows:

$$X++ \qquad X--$$

The postfix form means that the variable value is evaluated and used in the expression that it appears in; then the variable is incremented (++) or decremented ($--$) before execution proceeds. The usefulness of this option will be shown later during the discussions on program control flow in Chapter 4.

The unary **sizeof( )** operator is actually a function of the C language and is, in fact, the only function in the language (other functions are available in C, but they are all user-defined). The *sizeof( )* function returns the number of bytes that a data type, variable, or constant uses in the program. For example, Program 3.4 will tell you how many bytes each of the fundamental data types are allocated for your compiler. The output of Program 3.4 will vary, depending on the compiler, which is restricted by the architecture of the computer that it is run on. Try the program with your compiler and see what sort of data sizes that you have to work with. The **sizeof( )** function is very useful when working with arrays and structures and when using dynamic memory allocation. This function is critical when writing code that is sensitive to data type size. If extended precision for calculations is necessary, your program can use *sizeof( )* to test the hardware to verify that adequate variable storage size is present.

```
main(){
    printf("int:        %d bytes\n",sizeof(int));
    printf("short int:  %d bytes\n",
                            sizeof(short int));
    printf("long int:   %d bytes\n",sizeof(long int));
    printf("float:      %d bytes\n",sizeof(float));
    printf("double:     %d bytes\n",sizeof(double));
    printf("long double: %d bytes\n",
                            sizeof(long double));
    printf("char is:    %d bytes\n",sizeof(char));
}
```

Program 3.4   Use of *sizeof( )* function.

```
PROGRAM, SUBROUTINE or FUNCTION statement
variable declarations
EXTERNAL statements
DIMENSION statements
COMMON statements
EQUIVALENCE statements
DATA statements
Executable statements
END statement
```

Figure 3.3   Fortran program structure.

3.6   Program Structure, Statements, and Whitespace

In Chapter 2, we examined the C and Fortran program shells; now we will examine the basic program structure of the two languages in greater detail. Fortran program structure is shown in Figure 3.3, where the words in uppercase are reserved by the compiler. Typically, a simple Fortran program, as we have seen in Chapter 2, consists of a PROGRAM name statement, some variable declarations, program statements, and an END statement. Each line of a Fortran program is a program statement and must correspond to the column restrictions discussed in Section 2.5 unless the compiler has a special option to allow free-field input. In early Fortran compilers, blank lines were not permitted.

The C program structure is shown in Figure 3.4. The items in <<>> are optional. A minimal C program is just the **main( )** function followed by a pair of closed braces{ }. The C programs are free-field, which means that program elements can be anywhere in the file as

```
<< include files >>
<< defines >>
<< global declarations >>

main(){
<< external declarations >>
<< local declarations  >>
<< C program statements >>
}

<< functions >>
```

Figure 3.4 C program structure.

long as they are separated by whitespace (defined in greater detail in this section). A C program statement is identified by a terminating semicolon(;), and any expression may become a statement by the addition of a semicolon.

If the semicolon is omitted, errors can result. Examples of C statements are as follows:

```
++i;
x -= 10.0;
printf("Hello World!\n");
;
```

The last statement in the list above, a semicolon by itself, is the **null statement**. The computer execution of a null statement is to do nothing, which can be useful in some control structures that we will examine in the next chapter.

We can cluster C statements into what is called a **compound statement** by enclosing the statements with braces ({ }). The *main* function is just a single compound statement. The compound statement does not need to be terminated by a semicolon, although all statements within the compound statement must be terminated.

Whitespace is any of the following characters or sequences of these characters:

blank or space
tab
newline
carriage return
vertical tab
formfeed

Whitespace is ignored by the C compiler and does not print or display; however, whitespace characters add structure and readability to a program because the effects of these characters will be seen in displayed or printed output.

3.7   Formatted Output

At this juncture of our study, it will be useful to define output functions so that the results of our programs can be displayed. The primary C function for output is **printf( )**, and for Fortran it is **WRITE**. The *printf* function is part of the so-called "standard I/O library," and all C compilers include this function, although *printf* is <u>not a reserved word</u>. In Fortran, WRITE is a reserved word and is part of the compiler.

Formatted output means that the values to be outputted must conform to a format specified by the user. For example, if you have a floating point result that you want to output at a fixed precision, you can specify the desired precision to the output function (*printf* or WRITE). Internally, the value computed may be good to ten places, but you may only want to see five places after the decimal. Formatted output allows you to control how the data are presented.

In Fortran, the WRITE statement makes use of a FORMAT statement that specifies what the format of the data to be output is to be. The syntax of the WRITE/FORMAT pair is shown in Figure 3.5. We will go no further with the Fortran format specifications, although they are very similar to those of *printf*, which we discuss in detail below. You should note that the list of variables (X, Y, and Z in the figure) have a one-to-one correspondence with the format specifiers F4.3, F4.3, and F10.3. The F indicates that the variable value is to be output in floating point format. These specifiers cause the values to be outputted with field widths of 4 and 10, respectively, and a precision of 3 digits. The field width is the number of character spaces each output will have, and the precision is the quantity of digits that will appear after the decimal point. The $1X$ and $2X$ indicate the number of spaces between values output. For historical reasons, the first format specifier must be a $1X$, for this position was originally used for a printer control code.

The output *device unit number* determines which device will receive the data. If this is an asterisk (as in the figure), the data will be sent to the console or display. If the format statement number is also

3.7 Formatted Output

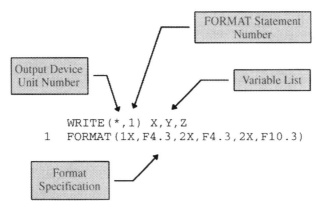

Figure 3.5 WRITE and FORMAT statements.

an asterisk, the data will be unformatted and outputted to the device using the default format specification for the data type. In our examples in this book we will use unformatted Fortran *WRITE* statements. When translating Fortran, it will be clear what values are being outputted (or inputted), and you should code the C output accordingly and not rely on a literal interpretation of the Fortran I/O statements.

The basic format for a *printf* statement to output an **ASCII** string is as follows:

```
printf("Hello World!\n");
```

All the characters within the quotes will be sent to the console (printed to the screen). The backslash-n (\n) is called an **escape character** and causes *printf* to output a **newline**. The *newline* is a combination carriage return and linefeed, such as on a typewriter. This just causes the screen cursor to move to the next line down. If you omit the newline, the data output by the next *printf* will start just after the exclamation point. For example,

```
printf("Hello World!");
printf("How are you?\n");
```

will output

```
Hello World! How are you?
```

Several clever outputs can be achieved using just *printf* and various escape characters. For example, the tab (\t) character can be used to align tabular data. If your program needs to alert the user to

Table 3.8 *Common Escape Characters.*

Name	Character	Effect
Newline	\ n	Carriage return/linefeed
Backspace	\ b	backspace
Tab	\ t	Moves cursor n* spaces
Alert	\ a	Sounds console beep
Vertical tab	\ v	Moves up one line (linefeed)
Carriage return	\ r	Returns cursor to beginning of line

*The variable n is operating-system-dependent.

a condition, when you output the alert (\a) character the console bell or beeper will sound. Table 3.8 lists the more common escape characters that can be used with *printf*.

If you need to output either a double quote (") or a back slash(\), then you must precede these characters with a backslash(\). This is called *escaping* the character. For example,

```
printf("\"Hello World!\"\n");
```

outputs a quoted "Hello World!", whereas

```
printf("It's either\\or.\n");
```

outputs It's either\or.

The quoted characters inside the parentheses of the printf are called the *format string*. If we want to output the values of variables, we must include special formatting characters within this string. Consider the printf statement below and assume that the variables $i = 1$, $j = 2$, and $k = 3$ have been declared integers:

```
printf("i = %d   j = %d   k = %d\n",i,j,k);
```

This printf will output:

```
i = 1  j = 2  k = 3
```

The variables whose values are to be output are placed following the control string and separated by commas. The **%d** terms in the control string are formatting characters that indicate that integer values are to be printed. Note the one-to-one correspondence between the three formatting characters (**%d**) and the variables. The percent symbol (%) is used by *printf* to determine how a variable is to be outputted. The symbol must be followed by a conversion character

Table 3.9 *Printf Conversion Characters.*

d, i	integer
c	character
f	float or double
e	double with exponential
%	no conversion, output % sign

that indicates how the variable value is to be interpreted. Table 3.9 lists a subset of conversion characters recognized by *printf* that will be useful to you in your programming:

The output produced by *printf* can be very confusing; therefore, a set of examples of different outputs of the same variables will be shown followed by a brief explanation. Assume the following variable declarations:

```
int i = 5;
c = 'A';
float x = 3.1416;
double z = 6.02e21;
```

Examples (*printf* with formats and output in boldface):

```
printf("i = %i\n",i);
i = 5
```

This is the simplest integer output *printf.*

```
printf("c = %c   c = %d\n",c,c);
c = A c = 65
```

The character variable **c** is outputted twice: first with character formatting (%c), thus outputting an *A*. The second format is integer (%d). Why is the output a 65? Hint: Check the ASCII table in Appendix C (decimal) for the value of *A*.

```
printf("x = %f   x = %d\n",x,x);
x = 3.141600 x = 0
```

65

The floating point variable *x* is outputted twice (first with float formatting (%f); thus the expected 3.141600).The second format outputs *x* as an integer (%d), yielding an unpredictable value.

```
printf("z=%f\nz=%e\n",z,z);
z = 6020000000000000000000.000000
z = 6.02000e+21
```

The double variable *z* is outputted twice: first with float formatting (%f), thus outputting the full precision of the mantissa. The second format is in scientific notation (%e). Note that the default precision is six digits (see Section 3.8).

```
printf("The rate was 5%%!\n");
The rate was 5%!
```

A string containing a percent sign (%) is output.

The variable conversion specification begins with the percent symbol and ends with a conversion character, as listed above. Further output formatting codes can be included between these. We will discuss formatting options for floating point output, which take the following form:

```
%w.pf or %w.pe
```

The **w** is a number indicating the field width or total digits to allow in the outputting of a value. The **p** is a number that determines the **precision** or number of digits to allow in the mantissa of the output. Usage is best shown by examples, and those that follow are based on the following expressions:

```
float pi = 3.141592654;
float f_pi;
double d_pi;

f_pi = 22.0/7.0;
d_pi = 22.0/7.0;
```

The first column of Table 3.10 lists various *printf* statements, and the second column shows the corresponding outputs. The float

Table 3.10 *Printf Example Outputs.*

`printf("%f\n",pi);`	3.141593
`printf("%f\n",c_pi);`	3.142857
`printf("%f\n",d_pi);`	3.142857
`printf("%e\n",pi);`	3.14159e+00
`printf("%e\n",c_pi);`	3.14286e+00
`printf("%e\n",d_pi);`	3.14286e+00
`printf("%3f\n",pi);`	3.141593
`printf("%3.2f\n",pi);`	3.14
`printf("%3.10f\n",pi);`	3.1415927410
`printf("%3.10f\n",c_pi);`	3.1428570747
`printf("%3.10f\n",d_pi);`	3.1428571429
`printf("%.3e\n",d_pi);`	3.14e+00

variable *pi* has been initialized to a calculator value, whereas *f_pi* and *d_pi* are calculated by the computer. Note that in some cases the outputs are identical despite different formatting. In the example

```
printf("%3.10f\n",pi);
```

the precision of 10 is greater than the available precision, which is 9. Note that the output is padded with a zero. Engineering conclusions based on this output could lead to errors. The *width* specifier may also be used with integer formatting codes to fix the number of places allowed for the value to be output in. This can be useful when outputting tabular data.

3.8   Formatted Input

Input in C is accomplished with the **scanf()** function, a cousin to *printf* that is also part of the standard I/O library. Input in Fortran is handled by the **READ** function, and like WRITE, READ is a reserved word for the compiler. The structure of READ and *scanf* mirrors their input counterparts.

Formatted input, like output, means that the values to be read must conform to a format specified by the user. In Fortran, the READ statement makes use of a FORMAT statement that specifies what the format of the data to be input is to be. The syntax of a READ/FORMAT pair is shown in Figure 3.6. We will go no further with the Fortran

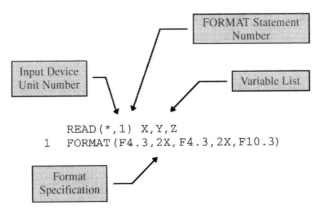

Figure 3.6 Fortan READ/FORMAT pair.

format specifications, although they are very similar to those of *scanf* as well as the WRITE statement. Unformatted READ is far more common. An asterisk replaces the FORMAT statement number, and the FORMAT statement is not used. The unformatted READ depends on the computer to format the data according to type.

With C, the basic format for a *scanf* statement to input data to variables (i_num: integer, x_num:float) is as follows:

```
scanf("%d%f", &i_num, &x_num);
```

There are subtle differences between the *scanf* specification and that of *printf*. Because *scanf* fetches a value to assign to a variable, the *address* of the variable and not the value must be given to *scanf*. This is done by prefixing the variable name with the address operator, an ampersand (&). The use of this operator is discussed in greater detail in Chapters 5 and 6. For now, just be sure to precede your variables with it when using scanf.

✓ One of the most common programming errors in C is the failure to put the ampersand (&) in front of a *scanf* variable.

Table 3.11 lists a subset of conversion characters recognized by *scanf*. Note that a conversion for type *char* (character) is not listed. Although *scanf* can convert *char*, it is not recommended. It is better to use *getch( )*, which is discussed at the end of this section after the

3.8 Formatted Input

Table 3.11 *Scanf Conversion Characters.*

d, i	integer
ld	long integer
f	float
lf	double

```
/* d_scanf.c
   program to demonstrate scanf() */

main(){
    int i_num;
    float x_real;
    double x_lrg;

     /* prompt and input data */
    printf("Enter an integer:");
       scanf("%d",&i_num);
    printf("Enter a float:   ");
       scanf("%f",&x_real);
    printf("Enter a double:  ");
       scanf("%lf",&x_lrg);

    /* output data */
    printf("You entered--\n");
    printf("\tinteger:%d\n",i_num);
    printf("\tfloat:  %f\n",x_real);
    printf("\tdouble: %.9f\n",x_lrg);
}
```

Program 3.5 d_scanf.c.

scanf examples.

Examine Program 3.5, **d_scanf.c**, which demonstrates how to use *scanf*. Three variables are declared, an int (i_num), a float (x_real), and a double (x_lrg). *Scanf* is then used to input values for the variables, and then they are immediately output with *printf*. A set of runs follows the program listing to show what the input and output look like. The user input is shown in boldface.

```
RUN #1 d_scanf.c
Enter an integer:1
Enter a float:    1
Enter a double:   1
You entered--
        integer:1
        float:  1.000000
        double: 1.000000000
```

In Run No. 1, three ones were entered and three ones were output. Note that the default precision of six digits was outputted for the float, and the forced nine digits were outputted for the double.

```
RUN #2 d_scanf.c
Enter an integer:65536
Enter a float:    3.141592654
Enter a double:   3.141592654
You entered--
        integer:0
        float:  3.141593
        double: 3.141592654
```

For Run No. 2, the entry of 65,336 exceeds the capacity of the (with this compiler) 2-byte integer. Thus, a zero was outputted. In the case of the float and double entries, the value of PI was determined from a calculator. The internal representation for the float variable is indeterminate from the output. *Printf* rounded to six digits after the decimal. The value for double is as entered.

```
RUN #3 d_scanf.c
Enter an integer:123
Enter a float:    1.987e10
Enter a double:   1.987e10
You entered--
        integer:123
        float:  19869999104.000000
        double: 19870000000.000000000
```

For Run No. 3, a number in scientific notation has been entered that exceeds the capacity of the float variable. Note that the computer has added data to the number. The internal representation is indeterminate from this output. The double variable has outputted correctly.

✓ *scanf* can yield strange and unpredictable results if variable type is not matched correctly to a conversion specifier.

The **getch( )** function in the standard library can be used to input a single character from the keyboard. The usage of *getch* is as follows:

c = getch( );

When this statement is executed, the computer waits until a key is pressed. The ASCII value of the key pressed is then assigned to the character variable *c*. You can use *getch* without a variable to effect a program pause, as in the following statements:

```
printf("<<< Press any key to continue. >>>");
getch();
```

The computer will output the prompt and then wait until a key is pressed. This can be any key, spacebar, return, and so on. This feature is useful to prevent the output from scrolling off the screen when the program outputs a large amount of data.

3.9   Precedence Rules

Now that we know how to form expressions, we can discuss the precedence rules of operators in greater detail. These rules are complex and can cause errors in the evaluation of expressions. When operators share the same level of precedence, evaluation proceeds according to the associativity of the operators. Operator precedence is less complex for Fortran than it is for C because of the large number of operators in C and the way that C evaluates expressions. We will not review Fortran precedence except to note that the exponentiation operator (**) has the highest precedence of the arithmetic operators.

Table 3.12 *Precedence and Associativity Rules.*

Operators	Associativity
`( )`	left to right
`++ -- + - sizeof`*	left to right
`*/ %`	left to right
`+ -`	left to right
`< <= > >=`	left to right
`== !=`	left to right
`&&`	left to right
`||`	left to right
`= += -= *= /= %=`	right to left

*Unary operators.

Table 3.12 lists the precedence and associativity for the C operators we have discussed, and Appendix E contains the complete table for all C operators and for Fortran.

Recall the following example from Section 3.2:

$$num = 5 * 3 * 2 + 1 * 10;$$

The compiler will evaluate this to $num = 40$ because multiplication precedes addition. The desired value may have been $num = 450$, which is computed from

$$num = (5 * 3) * (2 + 1) * 10;$$

Now consider the following:

$num = 5 * 3 \% 2$ evaluates to 1
$num = (5 * 3) \% 2$ evaluates to 1
$num = 5 * (3 \% 2)$ evaluates to 5

The first expression is ambiguous, and we must consult the precedence chart to evaluate it. From the chart, multiplication (*) and modulus (%) are equal in precedence; however, they associate left to right, and the multiplication is therefore evaluated first. Hence, the first and second expressions are equivalent. The assignment operators, however, associate from right to left. Consider how the computer evaluates the following statement:

$$i = j = 3 * 2;$$

Will i be set to the value of j or to 6? Because assignment associates *right to left*, j will be assigned the value of 6; then i will be assigned the new value of j. Assignment can get complicated with statements such as

$$i \; = 2;$$
$$i *= 3 + 2;$$

Once again, because assignments associate right to left, the computer evaluates $3 + 2$; then it multiplies by the current value of i (2) and assigns i the final value of 10.

3.10 Summary

The types and operators of both the C and Fortran language have now been discussed. The formation of expressions and statements has also been covered, and we have seen how to use formatted input and output functions to produce programs capable of processing data and generating results. With the knowledge of data typing and the problem-solving method given in the previous chapter, you are now ready to begin writing programs that can produce viable data. The most important outcome of this chapter to the engineer is the understanding of data representation in the computer in the form of typing restrictions and definitions. These representations must be understood to guarantee the validity of the data produced by your programs, and you must always know what the computer is doing to the data. In Chapter 4 we will explore how to control the sequence of statement evaluation and data processing, which will open up the vast computational power of the computer to us.

REVIEW WORDS

ASCII
Boolean result
case sensitive

char
compound statement
divide-by-zero
double
escape character
evaluation
expression
float
getch
int
left-hand side
long
modulus
newline
null statement
operator
overflow
postfix
precedence
precision
printf
promotion
reserved word
right-hand side
scanf
short
sizeof
truncation
type
underflow
unsigned
whitespace

EXERCISES

1.   Assume the following declarations:

     int $a = 20, \quad b = 10, \quad c = 5;$

Exercises

Evaluate the C expressions:

a. $a/b + c$

b. $a \% b \% c$

c. $c - - b$

d. $c - - - b$

2. The transfer function for a first-order system is given by

$$F(s) = \frac{1}{s + 1}$$

and a corresponding C expression is

$F = 1/(s + 1);$

Give similar C expressions for the following transfer functions:

a. $F(s) = \dfrac{1}{s^2 + 6s + 34}$

b. $F(s) = \dfrac{s + 4}{s(s^2 + 2s + 10)}$

c. $F(s) = \dfrac{s^2}{(s + 1)(s^2 + 2s + 20)}$

3. Write a C program to compute the area, circumference, and diameter of a circle, given the radius. The user should input the radius as a floating point value.

4. Write a C program to calculate the mass of atoms and molecules, given their atomic weight. The formula for mass is

$$m = \frac{\text{Atomic Weight in grams}}{\text{Avogadro's Number}}$$

Avogadro's number $= 6.02472 \times 10^{23}$ molecules (g-mole)$^{-1}$. Compute the mass of a hydrogen atom (atomic weight 1.008 g) and that of an oxygen molecule (atomic weight 32). Output your results with two digits of precision.

5. Convert the Fortran program below, which calculates the position and velocity of a falling object, to C.

```
      PROGRAM PVCALC
C   PROGRAM COMPUTES POSITION (Y) AND VELOCITY (V)
C   OF A FALLING OBJECT AFTER 1, 2 AND 3 SECONDS.
```

```
REAL TIME, GRAV
REAL Y, V,V0
GRAV = -32.0
V0 = 0
TIME = 1
Y = V0*TIME + 0.5*GRAV*TIME**2
V = V0 + GRAV*TIME
WRITE(*,*) 'Time Position Velocity'
WRITE(*,*) TIME, Y,V
TIME = 2
V0 = V
Y = V0*TIME + 0.5*GRAV*TIME**2
V = V0 + GRAV*TIME
WRITE(*,*) TIME, Y,V
TIME = 3
V0 = V
Y = V0*TIME + 0.5*GRAV*TIME**2
V = V0 + GRAV*TIME
WRITE(*,*) TIME, Y,V
STOP
END
```

6. A temperature gauge in an automobile typically reads from 100 to 260°F, and the temperature sensor at the engine outputs a corresponding 5 to 10 volts. Write a C statement to compute temperature (T) given sensor voltage (V).

7. Evaluate the following C expressions for $X = 0$, $Y = 2$, and $Z = 3$:

```
(X >Y) && (Y<Z)
 (Y==Z) || (X<Y)
 (X<Y) && (Y<Z)
        (!X)
(! (X>Y) ) && (Y>=X)
```

8. Use the edit–compile–run cycle of your computer to execute Program 3.4 to determine the sizes of the variable types on your machine.

9. Write an expression that will be true if a floating point number has a whole part and false if it only has a fractional part.

4 Control Flow

The ability of the computer to evaluate an expression and take action based on that evaluation is the foundation of automated decision making and a fundamental aspect of machine intelligence – the study of how to make computers display intelligent behavior. Control flow in a program is tightly coupled to the interpretation of data in a problem. What is relatively simple for an engineer to do on a calculator may be quite complex to explain to the machine via the program. Alternatively, many tasks with complex repetitive sequences, such as matrix multiplication, are best done by computer, for it is very easy to lose one's place while entering data into a calculator! Flowcharts are one of the best ways to diagram control flow in an algorithm, and a flowchart makes subsequent program coding easier. To begin this part of our study, we examine the two fundamental control structures of computer languages: if-then statements and loops. For example, the statement

```
IF the temperature exceeds 200 degrees,
    THEN activate the cooling valve
```

is easily flowcharted as

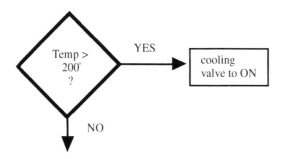

Our goal in this chapter is to learn how to complete the next step effectively, that of coding the control statements and using them in our programs. To do this, we will study the following C reserved words:

if	**break**
else	**continue**
for	**switch**
do	**case**
while	**default**
	goto

4.1 If

The basic if-then control structure is common to every computer programming language. Figure 4.1 shows the flowchart for the C **if** statement. The question mark inside the diamond is used to represent a C expression that is evaluated for a Boolean result. Recall that in C an expression that evaluates to zero is false, whereas an expression that evaluates to nonzero is true.

The syntax for the *if* statement is given by

```
if(<expression>) <statement>
```

If the expression in parentheses evaluates true (i.e., nonzero), then the statement immediately following is executed; otherwise, it is ignored. The statement to be executed can be a compound statement, and thus it is possible to have any number of statements executed based on the evaluation of the expression. To summarize, the if statement asks the question, Is the evaluation of the expression in parentheses true (nonzero)? If so, the statement immediately following is executed; if not, the statement is skipped.

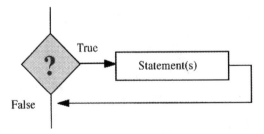

Figure 4.1 C **if** statement flowchart.

Let's apply our problem-solving steps and consider the calculation of the solutions to a quadratic equation. We need to solve $ax^2 + bx + c = 0$. That is our problem①. To do this, we need the values of coefficients a, b, and c; those are our inputs. We can then apply the well-known algebraic formula for the roots of a quadratic

$$x = \frac{-b \pm \sqrt{b^2 - 4ac}}{2a}$$

and get values for x, our output②. There are two potential problems in the use of this equation. The first occurs if the user should enter a zero for a, which would happen for a monic equation rather than a quadratic. If the user is neglectful and enters a zero for a, the computer will signal a *divide-by-zero* error and execution will cease. Consequently, we will want to check that the user inputs a valid quadratic (i.e., a $\neq$ 0). This is easy to do with an *if* statement by testing the expression a == 0. The expression is true if the value of a is 0, and thus at that point we can inform the user and exit the program.

Secondly, if the expression $b^2 - 4ac$ evaluates to negative, then the roots will be imaginary. The C math library contains a function, sqrt(x), that returns the square root of the argument x. The function will generate an error if the argument is less than zero. We will want to check for that condition and notify the user accordingly. We can do this by testing the expression $b^2 - 4ac < 0$, which is called the *determinant* of the quadratic. If true, we can either exit the program and inform the user that complex roots cannot be evaluated, or we can accommodate the complex results in our output.

For step three, work a sample set by hand, we can use the coefficients $b = 5$, $a = 1$, and $c = 1$, and our roots will be $x_1 = -4.791288$ and $x_2 = -0.208712$③. A pseudocode algorithm④ for this program is shown in Program 4.1. Note that we elected to constrain the user to real roots. We can now code the C program to compute the roots of a quadratic using *if* statements to avoid potential runtime errors⑤. The listing for this program is shown in Program 4.2.

After compilation, we run the program with our test case⑥ and get the following results (note that user input is shown in boldface type):

```
Enter coefficient a:1
Enter coefficient b:5
Enter coefficient c:1
```

```
Program quad_root
Declare input variables float a, b, c
Declare output variables float x1, x2
Declare test variable float D
Input coefficients a, b, c
If a = 0
    output "coefficients not quadratic"
    exit program
end if
D = (b*b) - (4*a*c)
If d < 0
    output "solution has complex roots
    exit program
end if
x1 = (-b - sqrt(D))/(2*a)
x2 = (-b + sqrt(D))/(2*a)
Output results x1, x2
exit program
```

Program 4.1   Pseudocode for quad_root program.

```
x1 = -4.791288
x2 = -0.208712
```

Because we want to make sure the *if* statements are working, we run the program and enter an illegal coefficient of $a = 0$:

```
Enter coefficient a:0
Illegal quadratic!
```

An entry of $a = 5$, $b = 1$, and $c = 1$ will produce a quadratic with complex roots; the program output shows that the *if* that tests for this case is working:

```
Enter coefficient a:5
Enter coefficient b:1
Enter coefficient c:1
complex roots!
```

Our program works, and we now have a quadratic root solver. Even though the program catches two potential problems, it is somewhat limited and fails to access much of the decision-processing power of the computer. Let's see how the use of an *if-else* statement can improve it. The C *if-else* allows for a selection between

4.1 If

```
/* Program quad_root.c */

#include <math.h>

main(){
    float a, b, c;
    float x1, x2;
    float D;

    printf("Enter coefficient a:");
    scanf("%f",&a);
    /* test for quadratic */
    if (a==0){
        printf("Illegal quadratic!\n");
        exit(0);
    }
    printf("Enter coefficient b:");
    scanf("%f",&b);
    printf("Enter coefficient c:");
    scanf("%f",&c);
    /* Compute argument for square root and
       test for complex roots */
    D = b*b - (4*a*c);
    if(D < 0){
        printf("complex roots! \n");
        exit(0);
    }
    /* Compute roots of quadratic */
    x1 = (-b - sqrt(D))/2*a);
    x2 = (-b + sqrt(D))/(2*a);
    /* Output the results */
    printf("x1 = %f\n", x1);
    printf("x2 = %f\n", x2);
    exit(0);
}
```

Program 4.2 C quadratic solver, quad_root.

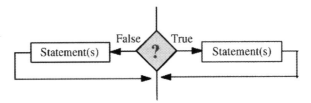

Figure 4.2 C **if-else** statement flowchart.

two alternatives, depending on the evaluation of the test expression. The syntax for the C *if-else* statement is given by

```
if(<expression>) <statement(true)>
    else <statement(false)>
```

If the expression in parentheses evaluates true (i.e., nonzero), then the statement immediately following is executed; otherwise, the statement following the *else* is executed. There is a subtle difference here between the *if* and the *if-else*, and the difference is best illustrated by an example as follows:

```
if(a==2) printf("a was equivalent to 2\n");
printf("a is equal to %d\n",a)
```

The boldface printf statement above will execute **if and only if** the value of *a* is equal to 2. With the *if-else* below, the output "a was equivalent to 2" will occur for a equals 2, and "a was not equivalent to 2" will output if *a* is not equal to 2.

```
if(a==2) printf("a was equivalent to 2\n");
else printf("a was not equivalent to 2\n");
printf("a is equal to %d\n", a)
```

In both cases the **printf("a is equal to %d\n",a)** statement will execute. The *if-else* is an "either-or" construct, as shown by the flowchart in Figure 4.2. Now let's see how the *if-else* can let us choose between alternatives to compute complex roots in our quadratic equation program.

✓ A common error when working with *if* statements is to forget that the statement following *if( )* is only executed when the expression evaluates true. If you need either-or, use *if-else*.

The quadratic solver (Programs 4.1 and 4.2) exited if the root argument was negative. This is because the root of a negative number is imaginary. We can change the sign of the argument and then change the output to reflect the complex root. Because a complex number is actually a vector in two-space, we need two more variables to hold the complex portion of the root. These can be declared as the float variables $x1i$ and $x2i$, the imaginary parts of $x1$ and $x2$. These changes are shown in Program 4.3. Note the control string of the two printf statements for the complex roots. The plus sign (+) preceding the float format character tells the computer to output the sign of the value, and this gives a straight column look to the output.

A run of the original Program 4.2 with the Program 4.3 changes is presented below. The determinant D will compute to zero, and the roots of the quadratic will be complex.

```
Enter coefficient a:5
Enter coefficient b:1
Enter coefficient c:1
x₁ = -0.100000-0.435890i
x₂ = +0.100000+0.435890i
```

If statements can also nest. This means that the statement following the *if* can be an *if*. Consider the following example:

```
if(a >= 0)
    if(c < 3)
        printf("a was >= 0 AND c was < 3");
```

The *printf* will only execute if <u>both</u> of the conditional expressions evaluate true. The second *if* statement is only tested if the first *if* statement evaluates true, and thus it would have been better to write the following:

```
if(a >= 0 && c < 3)
    printf("a was >= 0 AND c was < 3");
```

Fortran has two different *if* statements, the first being similar to the basic C *if*. The syntax of the Fortran *logical block* IF is shown by the following example translated from the quadratic test in Program 4.2:

```
IF(A.EQ.0)THEN
   WRITE(*,*)'Illegal quadratic!'
   STOP
ENDIF
```

```
/* Compute argument for square root
   process complex if negative */
D = b*b - (4*a*c);
if(D < 0){
    float x1i, x2i;
    D = -D;
    /* compute real part of roots */
    x1 = -b/(2*a);
    x2 = b/(2*a);
    /* compute imaginary part of roots */
    x1i = -sqrt(D)/(2*a);
    x2i = sqrt(D)/(2*a);

    /* Output complex results */
    printf("x1 = %+f%+fi\n", x1, x1i);
    printf("x2 = %+f%+fi\n",x2,x2i);
}
else {
    /* Compute roots of quadratic */
    x1 = (-b - sqrt(D))/(2*a);
    x2 = (-b + sqrt(D))/(2*a);

    /* Output real results */
    printf("x1 = %f\n",x1);
    printf("x2 = %f\n",x2);
}
exit(0);
```

Program 4.3 Changes to quadratic solver (quad_root) to handle complex roots.

Older Fortran compilers do not have the block IF form, and thus only one statement can be executed per IF. The other Fortran IF statement is called a computed or arithmetic IF. It has the following syntax:

IF(*expression*) <*S1*>, <*S2*>, <*S3*>

The *expression* must be arithmetic (i.e., compute to a number). If the value of the expression is less than zero, execution transfers to statement number <*S1*>; if equal to zero, execution transfers to

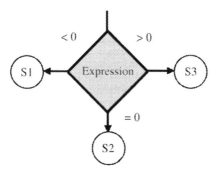

Figure 4.3 Fortran computed (arithmetic) IF.

statement number $<S2>$; and if greater than zero, to statement $<S3>$. The flowchart for the Fortran-computed IF statement is shown in Figure 4.3. The Fortran computed IF is well suited to test the quadratic determinant for our quadratic solver. When the determinant is less than zero, the roots are complex conjugate; when zero, they are real and equal; and when greater than zero, they are real and unequal. Program statements to process each of these cases could be numbered accordingly.

4.2 Loops

Loop structures allow the computer program to iterate or cycle over many values. This means that operations can be repeated, depending on a set range of values or on the truth or falsity of an expression. The C language permits three types of basic loops (we say basic because there are innumerable ways to combine loops):

1. **for loop:** used when a statement must be executed over a specific or predetermined range of values.
2. **do-while loop:** used when a statement must be executed and then a conditional expression evaluated to determine whether the loop should continue.
3. **while loop:** used when a conditional expression must be evaluated before loop statement execution.

All loops (independent of language) must have a **loop control variable**. This variable is initialized to a start value and is used by the loop to determine how many times to continue executing the loop statement(s). The loop statement must modify the loop control variable such that the loop halts at some point and program execution continues. If the loop control variable is missing, or if the loop statements fail to modify the loop control variable, the loop will continue until the program is halted by the operating system. This is what is called an **infinite loop** or a loop that does not halt. There are instances when an infinite loop is desirable, but for the most part infinite loops occur as programmer errors. It is important that you learn to identify the loop control variable.

To explore the use of loop statements, we will consider the problem of outputting a table of values. A Fahrenheit to Celsius (centigrade) temperature conversion table is a good choice to demonstrate the use of the C **for** loop. The problem is, Given a range of Fahrenheit temperatures and a delta temperature increment, what are the Celsius temperature values?① We need the formula for Fahrenheit to Celsius conversion, which can be quickly looked up in any physics text or engineering handbook. The formula is as follows:

$$°C = (5/9)(°F - 32).$$

We will need variables for the Celsius (C) and Fahrenheit (F) temperatures, which will also be our outputs②. Because we are given a range of values and an increment, these can be constants in the program, or we can allow the user to specify them. To simplify development, we will make the range from 32° to 212°F at 10°F increments and request no user input. For a temperature of 32°F, we expect 0°C, and for 212°F we should get 100°C ③. A flowchart for the problem is given in Figure 4.4④.

One might think that an *if* statement is called for in the coding of the algorithm because of the conditional block that evaluates the expression $F <= 212$. This is a common conceptual error. The *if* statement has no way to go back, or iterate, over prior statements in the program. We need to introduce a new mechanism that allows this to happen. One statement that can do this is the C **for** statement, with syntax

```
for(initialize; test; increment)<statement>
```

The C *for* statement has four parts: an *initialization* expression that is evaluated first to set up the loop control variable, a *test* expression that is evaluated to determine if execution of the loop should

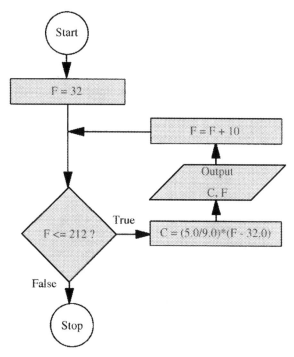

Figure 4.4 Flowchart for Fahrenheit-Celsius table generator with for loop.

continue, an *increment* expression to increment the loop control variable, and a *statement* (or compound statement) to be executed.

For our problem, we can use the Fahrenheit temperature variable *F* as the loop control variable. The C *for* expression for our program is

$$\text{for} (F = 32; \ F \ <= 212; \ F + = 10)$$

initialize ↑   ↑ test   ↑ increment

The sequence of evaluations is as follows:

1. The variable *F* is assigned the value 32 as a result of the initialization expression of the *for* loop.
2. The test expression *F* <= 212 is evaluated; if true, the statement following *for( )* is executed.
3. The expression *F* += 10 is evaluated, which increments the variable *F* by 10.
4. The cycle repeats steps 2 and 3 until the test evaluates false.

```
/* Fahrenheit-to-Celsius Table Generator */
main(){
    float F,C;
    for(F=32.0; F<= 212.0; F += 10.0){
        C = (5.0/9.0)*(F-32.0);
        printf("%3.0f\t%3.0f\n",F,C);
    }
}
```

Program 4.4   C Fahrenheit to Celsius table generator.

32	0
42	6
52	11
62	17
72	22
82	28
92	33
102	39
112	44
122	50
132	56
142	61
152	67
162	72
172	78
182	83
192	89
202	94
212	100

Figure 4.5   Fahrenheit to Celsius Table Generator output.

Now we can use the *for* statement to code the algorithm. The program listing is shown in Program 4.4⑤. Output from the program, two columns of temperatures, Fahrenheit and Celsius, is shown in Figure 4.5⑥. The output corresponds to our expectations for the temperature conversion.

Figure 4.6 is a flowchart for the C *for* statement. This figure introduces two added elements of all C loops, the **continue** statement and the **break** statement, which are C reserved words. When a *continue* statement is encountered during the execution of a loop, statements following the *continue* are ignored, and loop execution picks

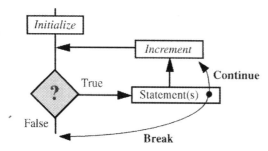

Figure 4.6 C for loop statement flowchart.

```
/* Fahrenheit-to-Celsius Table Generator
   with do-while loop */
main(){
   float F=32.0, C;
   do {
      C = (5.0/9.0)*(F-32.0);
      printf("%3.0f\t%3.0f\n",F,C);
      F += 10.0;
   } while (F<= 212.0);
}
```

Program 4.5   C Fahrenheit to Celsius table generator with **do-while** loop.

up at the next increment. A *break* statement causes the loop to terminate immediately. The *continue* and *break* statements are used with *if* statements that test whether to skip loop execution statements or to terminate the loop for some reason.

When algorithm requirements indicate that statements should be executed until a condition evaluates true, the C **do-while** loop structure is used. A *do-while* statement has the following syntax:

do <statement> **while**(<expression>);

The Fahrenheit to Celsius (centigrade) temperature conversion table could have been written with a do-while construct if the flowchart had been drawn, as in Figure 4.7. The modified program to use a *do-while* loop is shown in Program 4.5.

The flowchart for the C *do-while* (Fig. 4.8) is very similar to the *for* loop; however, be sure to note that the increment of the *loop control variable* must take place in the loop itself. This statement is

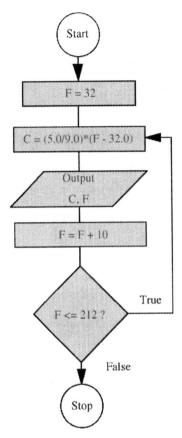

Figure 4.7 Flowchart for
Fahrenheit-Celsius table
generator with **do-while** loop.

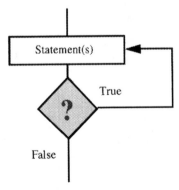

Figure 4.8 C do-while loop
statement flowchart.

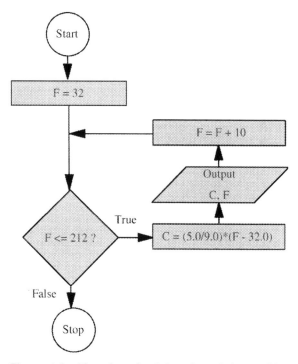

Figure 4.9 Flowchart for Fahrenheit-Celsius table generator with **while** loop.

useful when a set of loop statements must always be executed, although subsequent looping may not be needed, depending on what happens in the loop.

When the requirement arises to *test* and then *execute*, the C **while** loop is used. A *while* statement has the following syntax:

```
while(<expression>) <statement>;
```

The statement will repeatedly be executed as long as the expression evaluates true. Once again, the Fahrenheit to Celsius temperature conversion table could have been written with a *while* statement if the flowchart had been drawn as in Figure 4.9. The flowchart for the *while* statement is shown in Figure 4.10. Note that the *while* is identical in structure to the *for*; the initialization and increment statements, however, are not explicit. The modified program to use a *while* loop is shown in Program 4.6. Program output remains the same, as listed in Figure 4.5.

Although the same output can be achieved with all three loop forms, there generally is a "best choice" of loop structure. The criteria

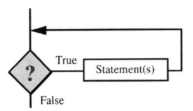

Figure 4.10 C **while** loop
statement flowchart.

```
/* Fahrenheit-to-Celsius Table Generator
   with while loop */
main()
{
    float F=32.0,C;
    while (F<= 212.0){
        C = (5.0/9.0)*(F-32.0);
        printf("%3.0f\t%3.0f\n",F,C);
        F += 10.0;
    }
}
```

Program 4.6 C Fahrenheit to Celsius table
generator with **while** loop.

for which loop to use are predicated on what you are trying to do.
The flowchart or pseudocode will indicate which loop to use. We will
now look at an example where a *do-while* is the loop of choice.

✓ When to use which loop?
If the loop is of fixed range, then **use a** *for* loop.
If statements must always be executed prior to testing the
loop control variable, then use the *do-while* loop.
If a test must be made to see if the loop should be entered
at all, then use the *while* loop.

The statistics program from Chapter 2 is an excellent example to
work from. The flowchart for this program indicates that a *do-while*
is the appropriate loop for the coding, even though a *for* loop was
used in the original example. A new coding using a *do-while* is given in

```
main(){
    int count= 10;
    float mean= 0.0, var= 0.0, dp, stddev;

    i=1;
    do{
    scanf("%f",&dpt);
    mean = mean + dpt;
        ++i;
    }while (i<=count);

    mean = mean/count;

    i=1;
    do{
    scanf("%f",&dpt);
    var = var +
        ((mean-dpt)*(mean-dpt));
        ++i;
    } while (i<=count);

    var = var/count;
    stddev = sqrt(var);

    printf("%f %f %f\n",
    mean,var,stddev);
}
```

Program 4.7 Statistics C code with **do-while**.

Program 4.7. The loop structure now matches the flowchart; however, we have to ask ourselves if this was really the best way to flowchart the problem. Because the problem statement specified exactly ten datapoints, a *for* structure would have been more appropriate even though the flowchart did not reflect this.

The question might be raised, What if we want to allow a variable number of values to be entered? One way to do this would be to ask the user to input the count at the start of the program. Another way would be through the use of a **flag variable**. A flag variable changes state when an event occurs and can be used to signal the program to do something. A flag variable can be used as a loop control variable, and an *if* statement can be used to detect when the flag variable state changes and thus causes an exit from the loop (using a break

```
main(){
    int nsamples = 0;
    float sample=0.0, mean = 0.0;

    printf("Enter data for mean\n");
    printf("(negative entry terminates)\n");
    while(1){
        printf("Input sample #%d:", nsample+1);
        scanf("%f",&sample);
        if(sample < 0)break;
        mean += sample;
        ++nsamples;
    }
    mean /= nsamples;
    printf("Mean of %d data values was %f\n",
            nsamples, mean);

}
```

Program 4.8   Infinite loop, flag variable, and **break** statement.

statement). For the statistics problem, we know that we will never have a negative value for the percentage of humans; therefore, we could have the user continue to enter values and signal the end of the data by entering a negative number. The data entry variable, sample, can be used as a *flag* variable whose state (sign) changes from positive to negative.

Program 4.8 illustrates the use of a *flag variable*, an *infinite loop*, and the *break* statement. In this program, the test expression for the *while* loop is a constant 1, meaning that it will always evaluate true and consequently the *while* will cycle indefinitely. After a sample value is inputted, we test to see if it is less than zero; if it is, then a *break* is used to "break out of the loop." A negative value for *sample* is the user's signal to indicate that data entry is complete. If the user enters a nonnegative value, that value is added to the *mean* variable, and the sample count variable *nsample* is incremented. Note that the user is informed how to terminate data entry and is prompted for each value. An example of the output of this program is shown in Figure 4.11.

In Fortran, the **DO** loop is similar to the C *for* loop. The syntax for a DO loop is as follows:

DO <statement#> <LCV> = <start>, <stop>, <inc>

```
Enter data for mean
(negative entry terminates)
Input sample #1:23
Input sample #2:12
Input sample #3:34
Input sample #4:29
Input sample #5:-1
Mean of 4 data values was 24.500000
```

Figure 4.11 Output example for Program 4.7.

```
PROGRAM F_TO_C
C Fahrenheit-to-Celsius Table Generator
C with DO loop
C
    REAL F,C

    DO 1 F=32.0,212.0,10.0
      C = (5.0/9.0)*(F-32.0)
1     WRITE(*,*)F,C
    STOP
    END
```

Program 4.9 Fortran Fahrenheit to Celsius table generator.

where `<statement#>` is the terminating statement of the loop, `<LCV>` is the loop control variable, `<start>` is the initial value of the loop control variable, `<stop>` is the final value, and `<inc>` is the increment. Program 4.9 is a listing of the Fortran version of the Fahrenheit to Celsius table generator program (see Program 4.4 for the C version). The output of this program is identical to that listed in Figure 4.5. A flowchart for the Fortran DO loop is shown in Figure 4.12. Note that the Fortran DO is identical in structure to that of the C *for* loop (Figure 4.6). The primary difference is that the start, stop, and end expressions are restrictive. Fortran expects numerical values for the start, stop, and increment (unlike the C *for*, they are not expressions to be evaluated). The increment value is optional, and, when omitted, the compiler assumes a value of 1.

Fortran, like C, has a CONTINUE statement, although it is substantially different in usage. The Fortran CONTINUE is simply used as a loop terminator and has no other function. Program 4.9 could have used a CONTINUE statement to close the loop following the

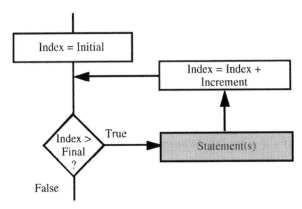

Figure 4.12 Fortran **Do** loop statement flowchart.

WRITE statement. In this case, the loop would have been as follows:

```
    DO 1 F=32.0,212.0,10.0
       C = (5.0/9.0)*(F-32.0)
       WRITE(*,*)F,C
  1 CONTINUE
```

4.3   Conditional Decision Structures

Conditional decision structures permit complex sequential evaluation of an expression against multiple target values. In C, the **switch**
statement yields the equivalent of multiple *if-else* statements in a
compact and very readable form. The flowchart for the C switch is
shown in Figure 4.13. There is no equivalent structure in Fortran.
The switch statement has the following syntax:

```
switch(<expr>){
   case <constant expression 1>: <statement>
   case <constant expression 2>: <statement>
                       ⋮
   case <constant expression N>: <statement>
   }
```

The N **case** statements are optional. If the case statements are
omitted, the switch reduces to an *if* where the first statement following the *switch* is executed if the *<expression>* evaluates true. This
usage renders the switch somewhat pointless, because the real power
of the *switch* comes from judicious use of the *case* statements.

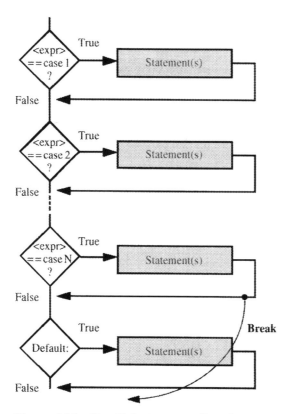

Figure 4.13 C **switch** statement flowchart.

From the flowchart (Figure 4.13), the result of the expression evaluation in parentheses following the reserved word *switch* is tested for equivalence in each *case*. If the expression is equivalent to the constant expression for the *case*, then the subsequent statement(s) are executed. Each *case* is evaluated in turn. If a *break* statement is encountered, then the *switch* structure is exited immediately with no further evaluations. A special *case*, the **default** case, is optional. The statements following the *default case* are always executed (unless the switch was exited by a break).

A good example of a *switch* implementation is that of an interface program for a radio-controlled car. Assume that a set of functions exists that when executed will perform the following actions:

```
forward()    --car moves forward.
stop()       --car stops moving.
left()       --wheels turn left.
right()      --wheels turn right.
backward()   --car moves backward.
```

```
while(1){
    printf("Enter command:");
    c = getch();
    if(c =='f')forward();
        else if(c =='s')stop();
            else if(c == 'l')left();
                else if(c =='r')right();
                    else if(c =='b')backward();
                        else if(c =='e'){
                            stop();
                            exit();
                        }
                        else
                    printf("Please enter f,s,l,r,b\
                    or 'e' to exit!\n");
}
```

Program 4.10   Control using **if-else**.

We want to write a program that will accept a user command in the form of a single character and call the necessary car control function. The user will input *f, s, l, r,* and *b* to control the car, or an *e* to exit the program. We will first examine the use of the *if-else* construct to execute this control, and this is shown in Program 4.10. The control is set up as an infinite loop, the *while(1)*, that continuously outputs the prompt "Enter command:" and waits for a single character input to the variable *c* from *getch( )*. The value entered is repeatedly tested through a set of nested *if-else* statements. If the value is an *f*, the routine to start forward motion is called. If the value is an *s*, the routine to stop the car is called, and so on. If no matching control character gets entered, the final *printf* will inform the user of the expected entries. Notice that if the user chooses to exit, the car is stopped before program termination.

Program 4.11 rewrites the control statements using a switch. Although at first glance this program may not appear any simpler than the *if-else* set, on examination it will prove to be the preferred solution to the problem. We have now eliminated the variable *c*. The switch can evaluate the *getch( )* directly. The character returned is compared for equivalence to each of the *case* statements. If none of the cases match, the *default* case will execute, and the user will be informed

```
while(1){
   printf("Enter command:");
   switch(getchar()){
      case 'f': forward(); break;
      case 's': stop();    break;
      case 'l': left();   break;
      case 'r': right(); break;
      case 'b': back();    break;
      case 'e':
                  stop();
                  exit();
      default:
         printf("Please enter f,s,l,r,b or\
               'e' to exit!\n");
   }
}
```

Program 4.11   Control using **switch**.

of the expected entries. Note that for each of the *cases* (except the exit and *default*) a break statement is added to stop any further evaluation within the switch after a successful match is made. Without the breaks, the switch would continue to test until the *default*, whose statement it would execute.

You may also employ multiple cases in your use of the switch. For example, if we had wanted to allow the user to enter either an *e or* a *q* (for quit) to exit, we could have written

```
case 'e':
case 'q':
         stop();
         exit();
```

This is the same as writing the following *if* statement from Program 4.10:

```
if(c =='e' || c =='q'){
      stop ();
      exit ();
}
```

The *switch* evaluates multiple *cases* as *or*.

4.4   Unconditional Control

The C **goto** and Fortran **GOTO** allow for unconditional program transfer. The syntax of the C *goto* and Fortran GOTO statements are

```
goto <statement label>  GOTO <statement#>
```

In C, the <statement label> is an identifier formed with the same rules as those for variable names. When the *goto* is executed, program execution transfers to the statement following the label. The label name must be followed by a colon (:) to distinguish it from a program statement. In Fortran, program execution is transferred to the statement <u>number</u> following the **GOTO**. An example of use of the C goto is shown in Figure 4.14.

It is not recommended that you use the C *goto*, for it conflicts with the structured nature of the language and makes programs difficult to follow. Nevertheless, in older Fortran programs it is used heavily, and you should therefore expect to encounter it.

✓ Should the C *goto* ever be used?

Yes, one example is very deep if-then or loop nesting and a condition arises in which the nest must be exited. Occasionally if-then nests will be patterned after some phenomenon or behavior, and proper form will not be possible; hence, the need for a *goto* escape. With loops the approved method of exit is the *break*, but this will only exit the loop where the break is encountered.

```
main()
{
        <program statements>
              ⋮
        goto my_label;
              ⋮
            <program statements>
              ⋮
        my_label:
            <program statements>
              ⋮
        exit(0);
}
```

Figure 4.14 C goto usage.

4.5 Summary

This chapter covered control structures in both C and Fortran. You should now be able to write fairly complex programs that produce useful data

REVIEW WORDS

break
case
continue
default
do-while
else
flag variable
for
goto
if
infinite loop
loop control variable
switch
while

EXERCISES

1. For the following C program statements, assume the following declarations:

```
unsigned char C;
int I;
float X;
```

In each instance, what is the output?

a.
```
C = 256;
if(C)printf("true");
  else printf("false");
```

b.
```
I = 256;
if(I%2)printf("Even?");
  else printf("Odd?");
```

c.

```
X = 0.0;
if(!X)printf("Yep!");
printf("Gotcha!");
```

d.

```
I = 1;
C = 2;
X = 3;
if(I<C && X>C)printf("true");
printf("or false?");
```

e.

```
C = -2;
if(C)printf("negative");
    else printf("or positive?");
```

2. Change Program 4.2 in such a way that if the user enters a zero for variable *a*, the program outputs the root of the monic equation instead of generating an error message.

3. Write a temperature conversion table generator program that allows the user to specify, start, stop, and increment temperature. The program should convert Fahrenheit to Celsius *and* Kelvin. A label indicating °F, °C, and K temperature should be outputted at the top of the table.

4. For the following C program statements, assume the declarations:

```
unsigned char C;
int I;
float X;
```

In each instance, what is the output?

a.

```
C = 5;
while(C!=0)printf("%d",C--);
```

b.

```
I = 10;
do if(I%2)printf("ODD");
      while(I--);
```

c.

```
for(C=-2; C<0;++C)
      printf("%d",C);
```

d.
```
for(C='A';  C<='Z';  ++C)
        printf("%C",C);
```
e.
```
for(X=0;  X==0;)
        printf("%f",X++);
```

5. Write a simple calculator program in C with the following features:

    *   User inputs any of the following commands:
        H -- program outputs a help message explaining commands.

        X -- program prompts for a float value for
                 variable X.
        Y -- program prompts for a float value for
                 variable Y.
        + -- program computes X = X + Y, outputs X
        - -- program computes X = X - Y, outputs X
        * -- program computes X = X * Y, outputs X
        / -- program computes X = X / Y, outputs X
        O -- program outputs current values of X & Y
        Q -- program exits

    *   Error message is generated for divide-by-zero
        request.

    Use the problem-solving steps and generate either a flowchart or pseudocode for your program.

6. From statics, a *linear force system* is as diagrammed in Figure A below. The equivalent resultant force F_R is equal to the sum of individual forces and is given by

    $$F_R = \sum_i F_i.$$

    The location of the centroid of this force, xR, is determined from the sum of the component force centroids, as follows:

    $$x_R = \frac{\sum_i F_i x_i}{\sum_i F_i}.$$

    Write a C program to compute any number of resultant forces and centroids for systems of this type. Consider only positive forces and distances. The user will flag completion of data entry with a negative value. Test for zero force before calculating the

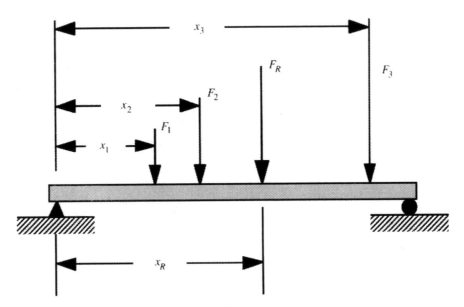

Figure A. Linear Force System.

centroid to avoid *divide-by-zero* errors. Use the problem-solving steps and generate either a flowchart or pseudocode for your program.

7. The voltage across the capacitor in a first-order RC circuit is given by the following expression:

$$v_C = I_S R + (V_0 - I_S R)e^{-t/RC}, t \geq 0.$$

Write a C program that outputs a table of values of t and v_C. The user enters the range of time desired in seconds and the increment in fractions of a second, the source current (I_S) in amperes, the resistance (R) in ohms, the initial voltage on the capacitor (V_0) in volts, and the capacitor value (C) in farads. Use the problem-solving steps and generate either a flowchart or pseudocode for your program.

8. The x- and y-coordinates of a ballistic projectile at any time (t) are given by

$$x = (v_0 \cos \theta_0)t$$

$$y = (v_0 \sin \theta_0)t - \frac{1}{2}gt^2$$

for some initial velocity v_0 and elevation angle θ_0. The acceleration due to gravity is g (32 ft s^{-1} or 975.36 m s^{-1}). Write a C program that allows the user to enter an initial velocity and angle, a time increment, and a stop time in seconds and that then outputs the x- and y-position of the projectile as a function of time.

9.   Modify the program of Exercise 8 to output the highest point that the projectile reached in flight and at what time it occurred. Hint: Use a variable to store the previous y value and an if statement to determine at what point it begins to decrease.

5 Type Conversion, Functions, and Scope

nterpretation of data is an important element in the process of describing how the computer acts upon data. If this interpretation is faulty, control structures will not operate properly or data will be misrepresented on output. The definitions of Fortran and C types were explained in Chapter 3; however, the underlying assumption was that types of variables would not be mixed. In other words, calculations involving real variables would only use real variables, and those involving integers would only use integers. Also, the size of variables, or how large a value a variable can hold, was not considered in any detail other than the observation that a double variable is twice the size of a float, and so forth. This chapter will explore the nature of type mixing and the importance of knowing the usage rules for type conversion.

As you know, a program is a set of instructions to perform a task. Large tasks can require programs of huge scope and size; thus, it is useful to be able to partition programs into logical segments. Subroutines or functions represent a useful way to achieve the partitioning of programs to add readability, manageability, and overall structure to a program. This is not all that functions allow us to achieve, for they also save the programmer from useless repetition of program code that is used often and in various places in a program. Finally, they allow for an elegant and efficient way to include mathematical functions in computer programs.

5.1 Casting and Type Conversion

The **cast** operator ( ) is used to specify explicitly the conversion of a variable or expression from one type to another. The syntax for a cast is

```
(type) <expression>
```

Recall the example from Chapter 3 in which type was mixed. The question was asked, What is the result of the following operation?

$$x = 5/2;$$

The solution was that x would be assigned the value of 2 because of integer division. We now add a fourth solution as follows that solves the problem of getting a floating point result:

$$x = (\text{float})5/2;$$

The preceding statement *casts* 5 to a float 5.0, and this value is divided by the integer 2. The result is a float 2.5. The *cast* operator causes a temporary type change that is in effect only during the evaluation of the statement in which it appears. Note that an alternate interpretation of the statement above might be that the cast takes effect *after* integer division. In other words, why doesn't the integer result of 5 divided by 2 get computed and *then* be cast to a float 2.0? The answer is that the cast operator has higher *precedence* than any of the arithmetic operators. It does not, however, have higher precedence than the parentheses used to force evaluation. Be careful of a statement such as

$$x = (\text{float})(5/2);$$

because x will evaluate to 2.0. The integer division in parentheses will have precedence over the cast.

When to cast depends on the type of the variables and constants used in the expressions being formulated. Rarely does the programmer specify type solely on anticipated operations. Rather, it is more likely that the variable data types will be determined by the nature of the data represented. This is particularly true of engineering problems. You should resist the urge to "just make everything double." This will lead to poor program design, and it will inevitably be difficult to follow what you are trying to do. Casting is very useful when mixing functions of different types and sending data to functions that are not of the functions type, and we will see how this affects our use and development of functions in Section 5.2.

Fortran has no facility for casting one type into another. The promotion rules for variables and the precedence rules for operators (described in Chapter 3) apply. When translating Fortran to C, promotion and precedence rules apply in the same way they did for C. Potential problems can arise when converting expressions containing the Fortran exponentiation operator (**). For example, the Fortran statements

```
REAL X,Y
INTEGER I,J

Y = 3
I = 5
J = 2

X = Y**(I/J)
WRITE(*,*)X
```

will produce an output value for X of 9.0, which is 3^2. The same will be true of the following C translation:

```
float X,Y
int I,J;

Y = 3;
I = 5;
J = 2;

X = pow(Y,I/J);
printf("%f\n",X);
```

The *pow()* function receives a double 2.0 as the result of the integer division of $I = 5$ and $J = 2$ being promoted to double *after* the values get to the function. The original Fortran statements are misleading, and it may be difficult to ascertain exactly the intent of the programmer. If the intent was to have Y raised to the 2.0 power as an outcome of the integer division, then fine. If not, either the I or J would have to be promoted to floating point. In Fortran this could have been done as follows:

$$X = Y**1.0*I/J$$

In C, we can use the cast operator:

$$X = pow(Y, (float)I/J);$$

> ✓ Type conversion problems can be difficult to spot. After making sure your algorithm is coded properly, look for a type conversion problem if output data do not match the expected output for your hand-worked example data set.

As a final example, consider Program 5.1 (output in boldface). A single float variable x has been declared and initially assigned a value of $5 + 1/3$. All of the constants on the right-hand side are integers. When the value for x is outputted with the first *printf*, we see a value of 5.000000. This output is consistent because the result of the integer division 1/3 is zero. The second assignment yields the expected algebraic result because the floating point division of 1.0/3.0 is indeed 0.333333. The third assignment includes a cast of *float*; however, the integer division inside the parentheses is still 0, and thus the cast does not give the desired result. The fourth assignment yields the desired output because the 1 is cast to float. When divided by an integer 3, the result is a floating point 0.333333.

```
main(){
        float x;

     x = 5 + 1/3;
     printf("%f\n",x);

     x = 5 + 1.0/3.0;
     printf("%f\n",x);

     x = 5 + (float)(1/3);
     printf("%f\n",x);

     x = 5 + (float)1/3;
     printf("%f\n",x);
}
5.000000
5.333333
5.000000
5.333333
```

Program 5.1 Casting example.

Our final topic is that of Fortran implicit types. In Fortran, any variable that has not been explicitly declared and begins with the letters *I*, *J*, *K*, *L*, or *M* is *implicitly* (automatically) typed as INTEGER. All other variables not declared are *implicitly* typed as REAL. Implicit declarations are superseded by explicit declarations. For example, REAL JBESSEL declares the variable JBESSEL as real even though it would be an implicit integer if left undeclared. The IMPLICIT statement in Fortran may be used to declare a set of variable names implicit. For example, the Fortran statement below causes all undeclared variables that start with the letters *W* through *Z* to be integers:

```
IMPLICIT INTEGER (W-Z)
```

✓ Implicit variables, although convenient in the short term, represent very poor program practice because errors that result from mistakes in variable type declaration are very difficult to detect. It is recommended that all variables be declared when writing Fortran programs.

5.2   Functions

The C programs that we have written until now have been coded as the **main function**. We have used this function exclusively to define our C programs, but in actuality a C program comprises one or more functions. **A function** in C is defined as a distinct program unit used to perform a specific task. All C programs must have a **main function** because it is defined as the **entry point** of the program or where things get started. Programs in C have access to **library functions** or functions that have been defined elsewhere and have been pre-compiled and stored in object form to be included with a program during the linking stage (see Chapter 1). The library functions that we have used include *printf, scanf, pow,* and *sqrt.* Section 5.3 of this chapter discusses more of the library functions commonly available to C programmers. Also discussed in that section are the Fortran intrinsic functions. For now, our interest is to generate **user-defined functions**, which are written by the programmer to include with the development of a program.

The syntax of a **function definition** in C is as follows:

```
<type> <label>(<parameter list>){ <statements> }
```

The *type* of the function is optional and will default to integer (**int**), although it is not good programming practice to omit the function type. The label is the function name that will be used when the function is called. **A parameter list**, enclosed in parentheses, lists the variable declarations that the function receives values of. If the function receives no variable values (like *getch()*), then the special **void** type should be used (*void* simply indicates that the object has no specific type). The *parameter (or argument) list* is followed by a set of function statements that are to be executed when the function is called. These statements must be enclosed in braces.

The best way to learn how to write functions is to create a few. We will start by building a function that is called for effect (i.e., a function that does not require any data and doesn't return any data). One such function would be a *beep()* function, which sounds the tone or bell on the terminal or system output device when called. To sound the terminal tone, we need to output an ASCII *bel* control code. This can be done with the escape code \a or by sending a decimal 7 to the terminal. This can easily be accomplished by using a *printf* to output the control code.

The beep function is listed as Function 5.1. The function is declared *void* because it returns no data. It also has a *void* argument list because it receives no data. The beep function can be useful to signal the user that a process has been completed. Each time the function is called, the terminal bell or tone is sounded once. To illustrate how to send data into a function, we can modify the beep function so that it beeps several times, depending on the value of an integer variable, *beeps*, that is determined by the user.

The modified beep function is listed in Function 5.2 (changes in boldface). Note that now we have included the variable declaration `int beeps` within the parameter list of the function. The value of the integer variable *beeps* is supplied by the user and is used for loop control of the *while* statement. The *parameter* list may have as many variables as are needed by the function; however, only variables that carry data *into* the function should be listed. If a variable is needed

```
/* beep -- sounds terminal tone or bell */
void beep(void){
   printf("\a");
}
```

Function 5.1 Beep function.

111

```
/* beep -- sounds terminal tone or bell */
void beep(int beeps){
   while(beeps--)printf("\a");
}
```

Function 5.2 Modified beep function.

```
/* beep -- sounds terminal tone or bell */
void beep(int beeps)
{
   int i;

   for(i=beeps; i>0;--i)printf("\a");
}
```

Function 5.3 Beep function with **for** loop.

within the function for some function-specific task, that variable can be declared within the function itself.

The *beep* function can be rewritten using a *for* loop and an auxiliary variable *i* as the loop-control variable. This change is shown in Function 5.3. This function has an advantage in that beeps must be positive or the function will return without action. If the *beep* function shown in Function 5.2 is sent a negative number, the function will beep until the negative values have cycled through to zero, which could be a substantial number of times.

As an example of function writing and the inclusion of functions in program statements, we will revisit the temperature table program of Chapter 4. The scale conversion formulas are

$$°C = (5/9)(°F - 32) \ and \ °F = (9/5)°C + 32.$$

We want two functions, *F_to_C* and *C_to_F*, that will, given a temperature, return the corresponding conversion. These functions are listed in Function 5.4 and introduce a new C reserved word, **return**. *Return* is used to terminate execution of a function and return execution to the calling function. If the return statement has an argument, the evaluation of the argument must match the type of the function and will be returned as the value of the function. In our temperature conversion examples, the functions return the result of applying the temperature conversion formulas. Function 5.4 shows

```
/* Fahrenheit-to-Celsius Table Generator */
main(){
   float F;
   for(F = 32.0; F <= 212.0; F += 10.0)
      printf("%3.0f\t%3.0f\n",F,FtoC(F));
}
```

Program 5.2   C Fahrenheit to Celsius table generator.

```
/* FtoC -- returns the conversion of
           temp in degress Fahrenheit
           to degrees Celsius */
float FtoC (float temp){
   return((5.0/9.0)*(temp - 32.0));
}
/* CtoF -- returns the conversion of
           temp in degrees Celsius
           to degrees Fahrenheit */
float CtoF (float temp){
   return(((9.0/5.0)*temp) - 32.0);
}
```

Function 5.4   Functions F_to_C and C_to_F.

Program 4.3 rewritten to include calls to F_to_C, which takes place within the *printf*. Now the value for variable C from before (previously computed in the loop with the formula) is returned by the function F_to_C. The output of this program remains unchanged. Note that the F_to_C function receives the value of *float* variable F, which takes values from 32.0 to 212.0 in increments of 10.0 degrees in the loop and returns the converted *float* value in Celsius degrees. The returned value is used by *printf* to output the table. Likewise, it is just as simple to do the opposite (i.e., output a Celsius to Fahrenheit table). This is accomplished by Program 5.3 using the C_to_F function. Note the change in range $(0 \rightarrow 100)$.

A function may only return a single value, but many values may be sent to a function in the parameter list. As an example of this, we design a function to compute *centripetal force*. Figure 5.1 illustrates the vectors involved. A mass m (kg), is constrained to move in a circle of radius r (m) at a linear speed v(m/s). The force F in newtons applied

```
/* Celsius-to-Fahrenheit Table Generator */
main(){
    float C;
    for(C = 0.0; C<= 100.0; C += 10.0)
        printf("%3.0f\t%3.0f\n",C,CtoF(C));
}
```

Program 5.3   C Celsius to Fahrenheit table generator.

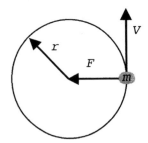

Figure 5.1   Centripetal
force system.

```
/* centripetalF -- returns centripetal force
              in newtons */
float centripetalF(float m, float v, float r)
{
    return(m*v*v/r);
}
```

Function 5.5   Centripetal force function.

to it is given by the following formula:

$$F = \frac{mv^2}{r}.$$

The function centripetal_F to compute force is shown in Function 5.5. Note that each variable in the *parameter list* must be declared uniquely and that all declarations must be separated with commas. This is because the parameter list is just that, a list of parameters (values) that the function is looking for; hence, each is declared uniquely

To see this, let us now consider a problem that would require the *centripetal_F* function. Centripetal force comes into play when a car uses a cloverleaf interchange to change direction while traveling on a highway. Given that the mean of a standard vehicle's weight is 2,000 lbs, the radius of curvature of a cloverleaf loop is 100 ft, and the speed limit is 55 mph, What are the forces acting on a car in the cloverleaf loop while speeds are changing from the speed limit to the ramp speed of 35 mph? Output these forces in increments of 1 mph.

We apply the problem-solving steps. The function *centripetal_F* will be called over the velocity range of 55 to 35 mph in increments of 1 mph to allow us to output a table of force values ①. Because the function *centripetal_F* requires data in mks units, we must first convert the given data into mks units. To do this, the following formulas are needed ②:

```
Mass in Kg = (Mass in Pounds)*0.45359
Velocity in m/s = (Velocity in mph)*.44704
Radius in meters = (Radius in feet)*0.30480
Force in pounds = (Force in newtons)*0.22481
```

At 50 mph, the car will be traveling at 22 m/s. The loop radius will be 30 m, and the mass of the car will be 907 Kg. The centripetal force will be 14,870 N or 3,343 lbs③. Quite a load on the tires and pavement! Now we can pseudocode the problem④:

```
Program cloverforce
Declare variables
   float carmass=2000
   carspeed=55
   loopradius=100
Define constants
   poundstokg = 0.45359237
   feettom      = 0.3048
   mphtomps     = 0.44704
   newttolbs  = 0.22480894
Declare output variable float force
carmass    = carmass * poundstokg
loopradius = loopradius * feettom
While carspeed >= 35
```

```
    force =
      centripetalF(carmass,
          carspeed*mphto mps,loopradius)
    output carspeed, force* newttolbs
    decrement carspeed
  end While
  exit program
```

The C code for this problem is shown in Program 5.4⑤. Note that the mks conversion constants were defined as program constants. Because the mass of the car and the loop radius do not change in the program, the data in these variables are converted to mks units first by multiplication with the conversion constants. The program then uses a *while* loop to cycle through the values of vehicle velocity given by the variable *car speed*, which is also used as the loop

```
/* cloverforce.c -- program outputs centripetal
forces
    acting on a 2000# vehicle in a 100 ft. radius
    circular ramp at speeds of 55 to 35 mph */

#include <stdio.h>

/* conversion constants used by program */
#define poundstokg 0.4536
#define feettom    0.3048
#define mphtomps   0.4470
#define newttolbs 0.2248

/* function to compute centripetal force */
float centripetalF(float mass, float velocity,
        float radius){
    return(mass*velocity*velocity/radius);
}
main(){
    float carmass=2000.0, carspeed=55.0,
    loopradius=100.0, force;

    /* convert fixed data to mks units */
    carmass *= poundstokg;
    loopradius *= feettom;
```

Program 5.4 Clover_force.c.

```
Speed    Force
  55     4044.13
  54     3898.41
  53     3755.36
  52     3614.98
  51     3477.28
  50     3342.25
  49     3209.90
  48     3080.22
  47     2953.22
  46     2828.88
  45     2707.23
  44     2588.24
  43     2471.93
  42     2358.29
  41     2247.33
  40     2139.04
  39     2033.43
  38     1930.49
  37     1830.22
  36     1732.62
  35     1637.70
```

Figure 5.2 Output of Clover_force.c program.

control variable. This variable is decremented after the current value is outputted by the *printf*. The *force* variable is assigned the return value of the function `centripetal_F`, which accepts the mass, velocity, and radius values. The value of the velocity variable *car_speed* is converted to before being passed to the function. The *force* variable value is converted to pounds prior to being passed to *printf*.

The output of the program is shown in Figure 5.2⑥. The neatness of the columns was achieved by forcing the output values to fixed widths in the *printf* statement. Note that the test value of 50 mph has been highlighted in boldface and shows agreement with our test data.

It is also possible for functions to call other functions, or for a function to call itself. This capability is important because it underscores that any statement possible in a program (until now, the *main function*) is also possible in a function. The C program is then just a collection of functions, beginning with the *main function*. The function simplifies and structures the program. A well-structured program is easier to follow in terms of the algorithm and also makes it easy to reuse functions from one program in another.

As we have seen, a function returns a single value depending on the type specified when the function is defined. If the function is not to return a value, then the *void* type is used. The question might now be, How do we get a function to return more than a single value? Recall that C passes the values of variables to functions. A function receives a copy of the value and cannot change the variable itself. For a function to be able to change the value of a variable, the *address* of the variable must be passed to the function. We have seen how an address is passed from our use of the address operator (&) in the *scanf* function. The function must be written to receive the address and work with it accordingly. This technique is beyond the scope of this chapter and requires knowledge of pointers, which we cover in Chapter 6; therefore, we will delay discussion of how to get multiple return values from functions until later.

Fortran makes use of both *functions*, which return values like their C counterparts, and **subroutines**, which are small programs that can be called like functions but have no return value. A C function that has been declared *void* is the equivalent of a Fortran subroutine. The most important difference between C and Fortran with respect to functions and subroutines is that Fortran passes data by reference (address), not by value. As discussed above, when you pass data to a C function, the data values are assigned to the variables that are local to the function: the ones defined in the functions parameter list. With Fortran, the address of the variables is passed to the function, and thus any assignments made to the variables in the function will be reflected in the variables declared in the calling routine.* This will be discussed in greater detail in Section 5.4, Data Scope. For now, we will just examine the basic syntax of the Fortran FUNCTION and SUBROUTINE.

The syntax of a function definition in Fortran is as follows:

```
<type> FUNCTION <label>(<parameter list>)
    <statements>
RETURN
```

The *type* of the function is optional and will default to the Fortran implicit type for the name. Although it is not good programming practice to omit the function type, untyped functions are often

---

*We use the word *routine* as a general term to refer to programs, functions, and subroutines.

5.2 Functions

```
REAL FUNCTION FTOC(TEMP)
FTOC = (5.0/9.0)*(TEMP - 32.0)
RETURN
```

Function 5.6 Fortran version of F_to_C function.

```
            SUBROUTINE FTOC(INTEMP,OUTEMP)
OUTEMP = (5.0/9.0)*(INTEMP - 32.0)
RETURN
```

Function 5.7 Fortran SUBROUTINE version of F_to_C function.

encountered when translating Fortran code. The RETURN statement terminates the function. The Fortran version of the Fahrenheit to Celsius conversion function is shown in Function 5.6. Note that the function name is used in the function itself to set the return value.

The only difference between a function and a subroutine is that the subroutine does not return a value. The Fortran FUNCTION is used in the same way as the C function as part of an expression. The subroutine, however, must be invoked with the Fortran reserved word **CALL**. This is the origin of the expression "call a subroutine." The syntax of a subroutine definition in Fortran is as follows:

SUBROUTINE <label>(<parameter list>)

<statements>

RETURN

The subroutine returns results as changed data values of the variables sent to it in the parameter list. Function 5.7 shows the C *F_to_C* function converted to a Fortran SUBROUTINE. The variable names INTEMP and OUTEMP are passed to the subroutine. The subroutine assigns the converted value of INTEMP to variable OUTEMP and returns. The calling routine variable that corresponds to OUTEMP has been assigned the new value. It is not necessary that the calling routine variable have the same name as that used in the subroutine, even though the value may be changed. The variables used in the subroutine definition parameter list are called *dummy variables* because they are only used as placeholders for the actual variables that reside in the calling routine.

> ✓ When should functions be written? The basic rules of thumb are to write functions when (a) the program uses the same set of statements repetitively in different sections of the program, making a single loop impractical, and (b) when a program calculation mirrors that of a physical process or mathematical formula.

5.3   Library Functions

Library functions are those functions available to the programmer from software libraries that are linked with the program after compilation. Software libraries are collections of precompiled object modules. You can generate your own libraries of functions that you define; the process for doing this, however, is beyond the scope of this book. Because C requires a **function prototype** (see Section 5.4, Scope), or definition, for all external function references, C libraries are accompanied by *include files* that define the prototypes for you. We have already used some of the functions defined in the **stdio.h** header file, which contains prototypes for the **standard i/o library**. Of primary interest to engineers are the **math library** functions. The **math.h** header file must be included when using functions from this library. A list of selected functions found in the math library is given in Table 5.1.

The **standard library** contains several important functions. The header file for the standard library is **stdlib.h**. Four functions are of interest to us from this library. The first is *int rand(void)*. This function returns a uniform random number between 0 and RAND_MAX, a system-defined integer constant. You can find the value defined for RAND_MAX by looking at the stdlib.h file. Typically, the value is 32,767. The random numbers generated are sequences derived from a polynomial formula. Eventually, the numbers repeat and are thus not truly random. For this reason, they are called **pseudorandom** numbers. The function *void srand(unsigned int seed)* allows you some control over the sequence, for the pass variable *seed* indexes the start position of the generator that *rand()* uses. The default value for the seed is 1, and each time you start a program that calls *rand()* the random numbers generated will be the same. If you set the seed value

Table 5.1 *Math Library Functions.*

`pow(double x, double y)`	raise x to the y power
`sqrt(x)`	square root of x
`log10(double x)`	logarithm base 10 of x
`log(double x)`	natural logarithm of x
`exp(double x)`	e raised to the x power
`fabs(double x)`	absolute value of x (float)
`sin(double x)`	sine of x, x in radians
`cos(double x)`	cosine of x, x in radians
`tan(double x)`	tangent of x, x in radians
`asin(double x)`	arcsine of x, x in radians
`acos(double x)`	arccosine of x, x in radians
`atan(double x)`	arctangent of x, x in radians
`sinh(double x)`	hyperbolic sine of x, x in radians
`cosh(double x)`	hyperbolic cosine of x (radians)
`tanh(double x)`	hyperbolic tangent of x (radians)
`ceil(double x)`	rounds x up $3.1416 \rightarrow 4.0$
`floor(double x)`	round x down $3.1416 \rightarrow 3.0$

to a different number each time, the sequence will change and be more random. There are various ways to do this. One is to ask the user to set the seed value; another is to use the *rand()* function itself to pick a seed.

The third function of interest out of the C math library is the *int abs(int i)* function, which returns the absolute value of an integer argument. We point this out because of the *fabs()* function in the math library (see Table 5.2), which returns the absolute value of a floating point argument. It is a common mistake to forget that *abs()* is not in the math library.

The last function of interest at this point is *void exit(int status)*. We have used this function to exit our programs. The function argument *status* can be used to send a program status code to the operating system. Some operating systems ignore the status values. The convention is to use a zero for a normal termination and a nonzero value when the program terminates because of a program-detected error.

Fortran incorporates what are called **intrinsic functions**, or functions that are "built in" to the language. There are over 100 intrinsic functions in Fortran, which are far too numerous to list here.

Table 5.2 *Fortran Intrinsic Functions and C Equivalents.*

Fortran Intrinsic	C
END	exit()
no intrinsic	rand()
no intrinsic	srand()
IABS()	abs()
X**Y	pow(X,Y)
SQRT()	sqrt()
LOG10()	log10()
LOG()	log()
EXP()	exp()
ABS()	fabs()
SIN()	sin()
COS()	cos()
TAN()	tan()
ASIN()	asin()
ACOS()	acos()
ATAN()	atan()
SINH()	sinh()
COSH()	cosh()
TANH()	tanh()
no intrinsic	ceil()
no intrinsic	floor()
MOD(I,J)	I%J

Table 5.2 shows the Fortran equivalencies of the C functions we have discussed. Always remember the fundamental difference between C and Fortran functions: C functions pass by value; Fortran functions pass by reference. In Chapter 6 we will see how variables can be modified in a C function through the use of pointers.

5.4   Data Scope

When variables and functions are declared, we have come to assume that they are available immediately to the *main function*, and they have been. In fact, if variables are not declared, the C compiler will produce an error. The range of use of a variable or function is what is called the **scope** of the variable or function. So far, we have only used variables that are **local** to the *main function*. Variable declarations have been placed at the beginning of the *main function*, and these variables have been available for use in our program. The C compiler has a "one-track mind" in the sense that it serially compiles a program file and it is not happy if it encounters symbols or objects that it does not understand. The compiler understands the structure of functions, reserved words, and expression–statement syntax, as well as anything that has been properly *declared*. The compiler keeps track of where things are declared, and this is what gives rise to *scope*, which refers to where in a program a variable or function declaration has meaning.

Because a C program is defined by the *main function*, the extent of a variable's scope can be defined in terms of *main*. This is illustrated in Figure 5.3, which graphically segments a C source file. Any variable declaration or function definitions that appear prior to *main* (the gray region at the top of the figure) will be global to all sections of the program that appear in the file. Those declarations that appear within the *main function*, the middle gray region, are *local* to *main*. Note that in C you may not define a function within a function, and thus only variable declarations and executable statements may appear in the center region. Any program elements declared or defined in the last region will be **external** to the *main function*. The reserved word **extern** is used when a variable is *external* in scope and must be made available either to *main* or another function.

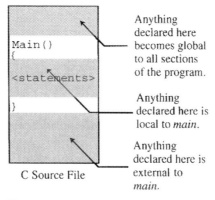

C Source File

Anything declared here becomes global to all sections of the program.

Anything declared here is local to *main*.

Anything declared here is external to *main*.

Figure 5.3 Scope of declarations to *main* function.

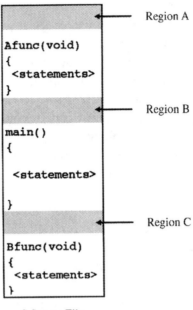

C Source File

Figure 5.4 Scope of declarations to *main* function and program elements.

Consider the three program regions A, B, and C illustrated in Figure 5.4. The source file consists of three function definitions: *Afunc*, *main*, and *Bfunc*. Variables declared within each of the function definitions will be *local* to those functions and available only to those functions. Any variables declared in Region A of Figure 5.4 will be *global* and available to any of the three functions. This can be advantageous and dangerous at the same time, for any of the functions in a program can change the value of a global variable.

Variables declared in Region B are *global* to *main* and *Bfunc* (which appear below them in the source) but are not available automatically to *Afunc*. Variables declared in Region C are *global* to *Bfunc* but are not available automatically to *main* or *Afunc*.

Consider Program 5.5. In this program, a variable *X* has been declared three times: once globally, once in the function *my_func*, and once in *main*. In each case, the variable has been initialized to a different value. When this program is run, it will output the following:

```
The value of X is 3.000000
The value of X is 2.000000
```

```
                                  ——— global variable
float X = 1.0; ↙
void my_func(void){       ———  local to my_func
      float X = 2.0; ↙
      printf("The value of X is %f\n",X);
}
                                  ——— local to main
main(){
      float X = 3.0; ↙
      printf("The value of X 0 is %f\n",X);
      my_func();
}
```

Program 5.5 Scope of same name variables.

```
                                  ——— global variable
float X = 1.0; ↙
void my_func(void){
      printf("The value of X is %f\n",X);
}

main(){
      printf("The value of X is %f\n",X);
      my_func();
}
```

Program 5.6 Scope of a global variable.

The first line of output corresponds to the *printf* in main and the value of X in main. The second line is the output of the *printf* in *my_func*, and it yields the value of the variable X local to *my_func*. The global variable X is superseded by the local definitions. This example should underscore the potential hazards of (1) using global variables and (2) using the same variable names across functions.

Program 5.6 is the same as Program 5.5 in which X has not been declared in either *my_func* or *main*. The global variable X is now accessible to *main* and *my_func*, and the output of the program becomes

```
The value of X is 1.000000
The value of X is 1.000000
```

Global variables are best used when a data element must be shared and changed by all of the components of a program. Often a *flag variable* (see Chapter 4) is best declared as a global. Recall that the flag variable indicates the state of something. One function can set the flag to indicate that data or a process has been completed, whereas others can read and reset the flag.

> ✓ When should global variables be used? The basic rule of thumb is to use them only when absolutely necessary. Such as with flag variables that are common to a large number of functions.

When a program variable is not within the scope of a particular function, the variable is *external* in *scope* to that function. For a function to use or have access to an external variable, the variable must be declared **extern**. Program 5.7 shows this with the variable *Y* declared after the main function. The *extern* reserved word gives accessibility to the external variable *Y* to both *main* and *my_func*. The output of this program will be

```
The value of Y is 1.000000
The value of Y is 1.000000
```

Function definitions are subject to scope rules just like variables. In Figure 5.4, the function *Afunc* is *global* to *main* and *Bfunc*, whereas

```
void my_func(void){
    extern float Y;
    pintf("The value of Y is %f\n",Y);
}

main(){
    extern float Y;
    printf("The value of Y is %f\n",Y);
    my_func();
}

float Y = 1.0;←——————  external variable
```

Program 5.7   Use of **extern**.

```
main(){                             function prototype
     extern float circumference(float r);

     printf("A circle of radius 5' has a\n")
     printf("circumference of %f'.\n",
          circumference(5.0));
}                              external function definition
float circumference(float r){
     return(3.1216*r*r);
}
```

Program 5.8 Use of **extern** function prototype.

Bfunc is *external* to both *main* and *Afunc*. If either *main* or *Afunc* need
to call *Bfunc*, a **function prototype**, which is a special term used to
describe a function declaration, must be supplied. Program 5.8 shows
a function circumference that is external to *main*. For *main* to access
a function circumference, we must supply the function prototype,
which is part of the declarations of the *main* function. Note that
the prototype looks exactly like the beginning of the function definition. The prototype tells the calling function (in this case *main*)
what the function returns as well as what the function expects in the
way of arguments.

There is one further caveat about the *function prototype*. The argument list does not need to have variable names that are the same
as those of the actual function because they serve only as placeholders. These variables are also called **dummy variables** because they
only indicate a variable type and not an actual variable. Hence, the
prototype could have been written as

```
extern float circumference(float x);
```

The variable *x* is not available to *main* because it has not been
declared. This is also true for the variable *r* in the function prototype
of Program 5.8. The variable *r* is available to the *circumference* function
because it is declared in the function definition. There is a subtle
difference between the function prototype, which indicates the type
of an external or global function and its argument types, and the
argument list types in the function definition.

Variables that are declared within functions other than **main**
are what we call **automatic** variables. An automatic variable only

```
main(){
  void how_many(void);
  int k;

  for(k=0;k<5;++k)how_many();
}                                    ————— automatic variable
void how_many(void){
  int I=1;
  printf("I've been called %d times. \n",I++);
  return;
}
```

Program 5.9 Automatic variable.

has value during the execution of the function where it is declared. The keyword **auto** is used to specify that a variable be placed in the *automatic* storage class, but this is redundant because all local variables default to *automatic*. The opposite of *auto* is **static**, and a variable declared *static* retains any value assigned to it for the duration of the program execution, regardless of where it is declared in a program. Variables declared in main are *static* by default, as are variables declared as globals. The *main* function in Program 5.9 calls a function, *how_many*, five times. Each time *how_many* is called, it outputs the value of a locally declared variable *I*. The value of this variable is then incremented before the functions return to *main*.

Output for this program is as follows:

```
I've been called 1 times.
I've been called 1 times.
I've been called 1 times.
I've been called 1 times.
I've been called 1 times.
```

This may not be what was expected. What is happening is that the variable *I* is reset each time we call the function because it is an *automatic* variable. To keep *I* from being reset, we need to declare it *static*, which is done in Program 5.10.

Output for the modified program is as follows:

```
I've been called 1 times.
I've been called 2 times.
I've been called 3 times.
I've been called 4 times.
```

```
main(){
  void how_many(void); int k;
  for(k=0; k<5; ++k)how_many();
}                                    static variable
void how_many(void){
  static int I=1;
  printf("I;ve been called %d times.\n",I++);
  return;
}
```

Program 5.10   Static variable.

```
I've been called 5 times.
```

This is what we wanted to see: the retention of the variable *I* value from call to call. The *main function* variables are *automatic,* and the reason they remain available throughout program execution is because we start from *main* to call all other functions and return to *main* before program termination.

✓ All variables are *automatic* variables unless they are declared globally (*extern*) or are specified as *static*. Use *static* to modify variable declarations in functions when you want the variable value to remain available throughout the execution of the program; otherwise, the function will reinitialize the variable value each time it is called.

One way to keep track of functions and variables in a program is through the use of a **program flow diagram**, or structure chart, which is just a flowchart that shows only the function calls and pass variables of a program. An example of a *program flow diagram* for a program that computes the height of tides at some time or when the tide will be at some height is shown in Figure 5.5. The pass variables have been left out to improve clarity. This program includes the following functions in addition to the *main*:

```
more         -- asks user to continue or terminate.
getParams    -- gets almanac data from the user.
TimeFloat    -- converts hours and minutes to minutes.
```

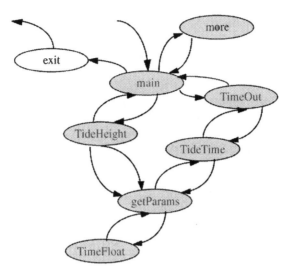

Figure 5.5 Program flow diagram for tides program.

```
TideHeight  -- computes height at some time.
TideTime    -- computes time at some height.
TimeOut     -- computes when tide will be out.
```

The diagram shows a single-ended arrow going into the *main* function, indicating where the program starts. The single-ended arrow leaving the *exit* function shows how the program terminates. Typically, library functions are not shown (in this program *printf* and *scanf* are used for I/O, but they have not been included on the diagram). The *exit* function is illustrated for clarity, but we could have omitted it and had the exit arrow leave *main* directly. From the diagram, *main* calls *more, TimeOut, TideHeight,* and *exit*. Each of these functions returns to *main* except *exit*, which terminates the program. *TimeOut* and *TideHeight* both call *getParams*, which calls *TimeFloat*. It should be clear from the diagram how the program flows with respect to functions called; however, no indication is given in the diagram regarding functionality of the functions.

Program flow diagrams are valuable when one constructs large software systems with many function calls, but the standard flowchart is adequate when writing small programs. We will return to the *tides* program in Chapter 6, for it illustrates the passing of data between functions.

5.5 Recursion

Recursion occurs when a function calls itself, and these functions typically arise in algorithms that generate sequences. The most commonly encountered recursive function in engineering is that of the **factorial**, and we will use it to illustrate how recursive functions work. A factorial is just the product of all the integers from one to a given integer. The mathematical symbol for the factorial is the exclamation point (!). As an example, the factorial of 8 is

$$8! = 1 \times 2 \times 3 \times 4 \times 5 \times 6 \times 7 \times 8 = 40,320$$

The factorial of a negative or floating point number is undefined, and the factorial of zero is, surprisingly, 1. Actually, this is not too surprising when one considers that the genesis and primary usage of the factorial are in the enumeration of things, and a zero can be considered a single item. Let's look at a function to compute the factorial using a loop:

```
/* compute factorial using loop */
int factoriall(int n){
  int temp=1;
  /* special case, zero */
  if(n == 0)return(1);
  /* countdown on f while multiplying
      the accumulated value, temp */
  while(n)temp *= n--;
  return(temp);
}
```

Now contrast the loop version with the following recursive function:

```
/* compute factorial using recursion */
int factorial(int n){
  /* special case, zero */
  if(n == 0)return(1);
  /* return the value n times the factorial of n-1
      -- the recursion of n! */
  return(n*factorial(n-1));
}
```

To understand the recursive version, consider $4! = 4*3*2*1$. When we call `factorial(5)`, it calls `factorial(4)`, which calls `factorial(3)`, which calls `factorial(2)`, which calls `factorial(1)`, which calls `factorial(0)`. This sequence occurs because on each call the function calls itself with $n - 1$. When the call to `factorial(0)` occurs, the function returns the value 1. The return sequence up the chain is then as follows:

```
factorial(0)  returns 1                          = 0!
factorial(1)  returns 1*factorial(0) = 1*1 =  1 = 1!
factorial(2)  returns 2*factorial(1) = 2*1 =  2 = 2!
factorial(3)  returns 3*factorial(2) = 3*2 =  6 = 3!
factorial(4)  returns 4*factorial(3) = 4*6 = 24 = 4!
```

A recursive function requires a **termination condition**, which is the recursive equivalent to the *loop control variable*. In the factorial function, the initial test for $n == 0$ is the termination condition. This expression causes the function to return without calling itself, and one needs to be sure that eventually the termination condition will occur. The *recursion* takes place in the second statement of the function, and the decrementing of $n(n - 1)$ guarantees that the termination will occur after n calls to the function. If the termination condition is missing, the function won't stop recursing, and an overflow exception will occur.

Recursion is a fundamental concept of mathematics and finds extensive use in computer science. Engineers tend to avoid recursion because of the initial abstraction of the concept. A recursive function will always be programmatically simpler than a loop-based equivalent and in some applications will be the strategy of choice.

REVIEW WORDS

auto
automatic
cast
dummy variables
entry point
extern
external
function
function definition
function prototype

implicit type
intrinsic function
library function
local
main function
math library
parameter list
program flow diagram
pseudorandom
recursion
register
return
scope
software libraries
standard library
standard I/O library
static
subroutine
termination condition
user-defined function

EXERCISES

1. Write a C function, *tconvert*, with the following definition:

   ```
   float tconvert(int towhat, float temp)
   ```

   If the variable *to_what* has a value of 1, the function should return the Fahrenheit conversion of the Celsius *temp*; if 2, the function should return the Celsius conversion of the Fahrenheit *temp*; and if 3, the function should return the Kelvin conversion of the Celsius *temp*.

2. The transcendental sine and cosine functions are easily represented and computed using the following series expressions:

$$\sin x = x - \frac{x^3}{!3} + \frac{x^5}{!5} - \frac{x^7}{!7} + \bullet \bullet \bullet$$

$$\cos x = 1 - \frac{x^2}{!2} + \frac{x^4}{!4} - \frac{x^6}{!6} + \bullet \bullet \bullet$$

   Write a C program that implements the expansions given above as functions. Name them Sin and Cos to distinguish them from the sin and cos functions in the math library. Also, write a factorial function that will be called by the Sin and Cos functions as needed.

Provide two arguments to Sin and Cos in which the value to be computed and the number of terms to be evaluated are in the expansion. You may use the *pow* function for the powers. If the number of terms is zero, then Sin(x) should return x, and Cos(x) should return 1.

Your main program should accept a value for x and the number of iterations. It should then output the value of sin(x) and cos(x) and then a table of three columns showing the iteration, Sin(x) and Cos(x). Contrast the values of your functions versus those in the math library.

3. The need to integrate functions often occurs in engineering problems. To integrate means to find the area under a curve, as shown in Figure E5.1. We can simplify the process by *sampling* the function at discrete intervals. Each sample has a trapezoidal shape from the line connecting the points along the function where the sample occurs, as shown in Figure E5.2. The *area* of a trapezoid

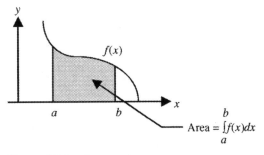

Figure E5.1   Area under a curve.

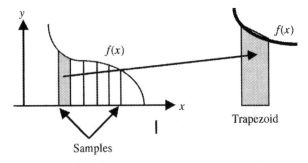

Figure E5.2   Trapezoid representing sample of area under curve.

is given by the following expression:

$$\text{area trapezoid} = \frac{1}{2}\text{base(height\_left + height\_right)}.$$

We can use this expression to compute all the sample trapezoid areas and then add them together to get the function area.

Write a C program to integrate over a range, at a user input increment value, a selected function using this trapezoid technique. The program should demonstrate integration of $\sin(x)$, $\cos(x)$, and x^2. Use a switch control structure in your program to select between the different functions.

4. The recursive Fibonacci number sequence is given by the following expression:

$$F_N = F_{N-1} + F_{N-2} \text{ for } N \geq 2 \text{ with } F_0 = F_1 = 1.$$

Write a recusive C function:

```
int fibonacci(int N)
```

that returns the Nth number of the sequence. Use the function in a program to output the first 15 numbers in the Fibonacci sequence.

5. Look for the float.h include file on the system that hosts the C compiler that you use. What is the value for FLT_MANT_DIG, the number of significant digits that a float variable can have? Write a C program that outputs successively smaller positive float values and determine when the limit is reached and what happens when it does.

6. For Exercise 5 of Chapter 4, add a set of functions to the calculator, such as $\sin(x)$, $\cos(x)$, $\tan(x)$, and $\text{sqrt}(x)$.

7. Write the trajectory program of Exercise 8 in Chapter 4 with functions that compute the x- and y-positions of the projectile.

6 Pointers, Arrays, and Structures

The C programming language has strong ties to assembly language, and this is evident in the direct access to memory features that the language includes in the form of pointers. Pointers allow a C program to have complete control over data type, data values, and data access. The engineer has a keen interest in pointers because they allow functions to access variables indirectly, and pointers form the basis of subscripted, or array, variable representations.

6.1 Pointers

A **pointer** is nothing more than a direct memory reference, or an address. A C *pointer* is a variable whose value is an address. This can be very confusing, for we have come to see variable names as being address references in themselves; however, there are times when we want to access a memory cell indirectly via the address of the cell. To do this we need a variable whose value points to a memory cell; hence, the term *pointer*. To declare a pointer variable, we use the unary pointer operator(*) in front of the variable name in the declaration. Pointers are sensitive to type because type determines how large a variable is in terms of number of bytes. To declare a pointer variable to an integer, we would use the following declaration:

```
int *pk;
```

The variable *pk* is called a *pointer* to an integer. We now want to assign to the variable *pk* the address of some integer variable because the value of a pointer is interpreted as an address. To do this we can use the unary address operator(&). Recall that all along we have used

```
#include <stdio.h>
main()
{
    int k,*pk;                          pointer
    pk = &k;                              to k

    k = 10;   integer k  ──────┐    │    ┌──────  value at
                               ↘   ↓   ↙          pointer
    printf("%d %ld %d\n",k,pk,*pk);
}
```

Program 6.1 Pointer.

this operator in *scanf* calls to signify the address of variables passed into *scanf*. Let's say that we have an integer variable *k*; then we can use the following assignment to assign the address of *k* to the pointer variable *pk*:

$$pk = \&k;$$

Now the value of *pk* is the address in memory of *k*. Consider Program 6.1 in which we have declared both *k* and *pk*, as discussed above. We've also assigned *k* and *pk* values: *k* the value of 10 and *pk* the value of *k*'s address. What gets outputted when the *printf* is executed?

The *printf* outputs values for *k*, *pk*, and **pk*. The output of this program is as follows:

```
10 2025066460 10
```

The first value should be understandable immediately because the program assigned *k* a value of 10 in the second statement. The second number may be a bit confusing because it is the address of the variable *k*, which is the value of the pointer variable *pk*. This number will change depending on the system and how the memory is allocated, and it is rare to actually output the address itself; we did so here for illustration. Note that we used %ld in the *printf* control string, the specifier for a *long int*. All pointers (addresses) are long integers, even though they may point to variable values of any type. The final value outputted is the value at the address that *pk* is pointing to, which is just the value of *k*. It should be clear that there are now two ways of representing a variable: the variable itself and the variable with a pointer.

Although there are several things that can be done with pointers, our interest is primarily in the passing of variable addresses to functions. Recall that in Chapter 5 we were restricted to functions that worked with variable values only because C passes only values of variables in the function argument list. To get data back from a function, we had to use the value of the function itself – a single value. If we use a variable address in the argument list of a function, the function has the ability to manipulate the variable value. This occurs with the *scanf* function when we pass the address of the variable(s) that we want input values returned for.

To illustrate the writing of a function that returns multiple values, consider the conversion of rectangular to polar coordinates. A coordinate pair (x,y) is converted to the equivalent polar coordinates (r, θ) by the following formulas:

$$r = \text{sqrt}(x^2 + y^2) \qquad \theta = \tan^{-1}(y/x)$$

Figure 6.1 shows a point at rectangular coordinates of $(1,1)$, which should convert to polar coordinates of $(1,45°)$. We want to write a function with arguments x, y, r, and theta. The variables x and y will be inputs, and the function will return values for r and theta based on the preceding formulas. A function to do this, rec_to_polar, is shown in Function 6.1. Note that the arguments in the function definition include the pointer operator $(*)$ on r and theta. This indicates that an address must be passed to the function (in this case, addresses for two float variables). In the body of the function, the statement

$$*r = \text{sqrt}(x*x + y*y);$$

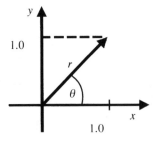

Figure 6.1 Rectangular and polar coordinates of a point.

```
void rec_to_polar(float x, float y,
                  float *r, float *theta)
{
    *r = sqrt (x*x + y*y);
    *theta = atan(y/x);
    return;
}
```

Function 6.1 Function to convert rectangular to polar coordinates.

```
#include <stdio.h>
#include <math.h>
main()
{
    extern void
      rec_to_polar(float x, float y,
                   float *r, float *theta);
    float r, theta;
    rec_to_polar(1.0,1.0, &r, &theta);
    printf("(1,1)->(%f,%f)\n",r,theta);
}
```

Program 6.2 Use of rec_to_polar function.

sets the *value* of r to the square root. If the pointer operator is omitted, the pointer itself will be set to the square root, and the value of the r variable in the main program will not be correct. Program 6.2 shows how we might use this function. In the program, two float variables, r and theta, have been declared. When we call the `rec_to_polar` function, the addresses of these variables are passed as the last two arguments. Because our example called for the conversion of rectangular coordinates (1,1) there was no need to declare variables for x and y. When we run the program, the following is outputted:

```
(1,1)->(1.414214,0.785398)
```

which is what we expected. Note that the angle is in radians. Also note that we included the header file for the math library (math.h) because our function uses *sqrt* and *atan*.

The following summarizes the use of pointers to write and use functions that modify variable values:

1. Decide what variables the function will modify. If the number of variables to be modified is one, then don't use pointers. Have the function return the value as itself.
2. Write the function using the pointer operator (∗) to indicate which variables will return values to the calling function.
3. Use the pointer operator (∗) on the variables to be modified in the function to cause their values to be assigned in the function.
4. Use the pointer operator (∗) to indicate the modifiable variables in the function prototype.
5. Use the address operator (&) when calling the function to indicate which variables will return values.

Because Fortran subroutines and functions pass the reference, or address, of variables (see Section 5.2), there is no use of pointers. Newer Fortran compilers (Fortran 90) include the ability to use pointers and memory references, but the implementation details are complex and beyond the scope of our discussion.

6.2   Arrays

Arrays in C are defined in several different ways. A C array is a pointer to a data space. The simplest representation of an array in C is done by declaring a variable and giving it dimension with bracket notation, as illustrated below:

```
float x[5][5], y[3], z[3][3][3];
char  s[10];
int   I[2][2];
```

Here, x is a 5×5 matrix of type float, y a float vector of length 3, and z is a 3-D matrix of type float. The variable s is a character array of 10 characters, and i is a 2×2 integer matrix. Array **indices** in C start at zero and go until the array size minus one. Consider the following integer array from before:

```
int I[2][2];
```

The values of this array are indexed at

```
I[0][0]  I[0][1]  I[1][0]  I[1][1]
```

in row-column order. Data in the array are accessed in a way similar to that used for other C data, except that now indices must be used to identify the individual variable elements of the array.

It is also possible to initialize the data in an array when it is declared. For example, the following code initializes a float array x to be a 1×6 vector with the values 1.0 through 6.0:

float $x[] = \{1.0, 2.0, 3.0, 4.0, 5.0, 6.0\}$;

Because arrays are data spaces, we cannot form arithmetic, logical, and relational expressions with them as collective groups. Array elements, however, are references to individual variables; therefore, we can use them in any expressions for which we would normally be able to use a common variable. We can also index arrays using integer variables. The following statements will multiply each element of our vector **x** (defined above) by 2.0, assuming that I has been declared *int*:

```
for(i=0;i<6;++i)x[i] *= 2.0;
```

We can output the result using the following:

```
for(i=0;i<6;++i) printf("x[%d]=%f\n",i,x[i]);
```

The output will be

```
x[0] = 2.000000
x[1] = 4.000000
x[2] = 6.000000
x[3] = 8.000000
x[4] = 10.000000
x[5] = 12.000000
```

Note the change in the index variable i. Caution must be observed when indexing array variables. You will note that i varies in the statements above from 0 to 5, the number of elements in **x**.

✓ A common error that occurs when using arrays is to confuse the number of elements declared with the index range. If you declare an array of dimension D, the index will vary from 0 to $D - 1$.

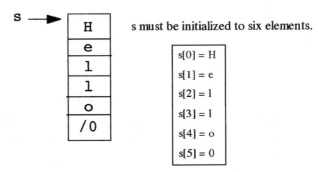

Figure 6.2   Character string memory allocation.

A character array is called a string in C and demands special considerations in usage. Consider the following character array declaration:

```
char s[10];
```

The declaration allocates space for 10 character elements; however, the 10th element is reserved for the **null character, 0. A string constant** is defined by quotation marks, for example,

```
char s[] = "Hello";
```

initializes the character array *s* to be 6 elements. The size of the character array was left blank, and the C compiler uses the string in quotes to determine how large a space to allocate. Additionally, the C compiler automatically includes the *null character*, or zero, for you. This placement of the null is illustrated in Figure 6.2. The variable *s* by itself is a *pointer* to the character string in memory. To access the individual characters of the string, we use the bracket notation (i.e., `s[0]` is the first element, `s[1]`, the second, etc.). The *null character* indicates the end of the string, which means that C strings can be of any length to the limit of memory. This is in contrast to Fortran strings, which have a limit on length.

If we execute the following C assignment,

```
s[1] = 'a';
```

s will become

```
s = "Hallo"
```

We could have used

```
s[1] = 0x61;
```

to do the same thing. Why?

✓ Remember that character array elements are individual single-byte characters. You can use them as one-byte integers, assign character constants (e.g., 'A') to them, or use ASCII codes.

Strings are arrays of characters; consequently, like numerical arrays, we cannot form arithmetic, logical, or relational expressions with them. We can, however, form expressions with the individual elements of strings. We can also use variables to index strings. For example, the following set of statements will output the contents of the string *s*:

```
i=0;
while(s[i])printf("%c",s[i++]);
printf("\n");
```

An easier way to output a string is to use the string format character in the *printf* control string, as follows:

```
printf("%s\n",s);
```

Just as in the example above with the *while* loop, the *printf* "looks for" the null character at the end of the string to tell it when to stop outputting characters. Several functions to manipulate strings are available in libraries that are typically supplied with C compilers. You can write your own functions to manipulate strings by operating on the individual elements and testing for the null character that terminates the string. Remember that the null character is a binary or integer zero, not the ASCII character zero, which has a value of 48. The string size is determined by the number of elements assigned in the declaration statement. If you manipulate the string in any way by moving the null character, you must remember not to index beyond the end of the string, or unpredictable results will occur.

Implicit Fortran arrays are defined by a DIMENSION statement, and explicit arrays are designated in the declaration of the variable. In the example below, **X** is a 5×5 matrix of implicit REAL, **Y** a REAL vector of length 3, **Z** is a 3-D REAL matrix, **S** is a character string of length 10, and I is a 2 × 2 integer matrix.

```
DIMENSION X(5,5), Y(3), Z(3,3,3)
CHARACTER S(10)
INTEGER I(2,2)
```

In the example above, the values of the *I* array are indexed at

```
I(1,1) I(1,2)
I(2,1) I(2,2)
```

in row-column order. The index range of an array may be changed using the following construct in the declaration:

```
INTEGER I(0:1,0:2)
```

Now the indexes will range from zero to 1, as with the C arrays. Other than determining the array type and the indexing range (be careful because the default is a start index of 1, not zero), the Fortran array is accessed and used in the same way as the C array. The Fortran string is used somewhat differently because the string has no termination character but is employed in the same way as an array. Newer Fortran compilers have string manipulation characteristics similar to C, but older Fortran code will have very few string manipulation operations, if any.

6.3 Structures

A C **structure** is a mechanism for representing groupings of differently typed variables. The most commonly used structures in C for engineering purposes are those for coordinate systems and for complex numbers. Using the example of complex numbers, one might wonder how a single representation could be made to accommodate both the real and imaginary parts. We can define a C *structure* that allows us to declare a single complex variable having two parts, or members, by using the C reserved word **struct**. The **structure definition** for a complex number could be as follows:

```
struct complex {
    float real, imag;
};
```

This definition defines a structure type called `complex` for which each structure variable declared `complex` has two members, a float member named `real`, and a float member named `imag`. Now we can declare **structure variables** based on the following definition:

```
struct complex a, b;
```

The preceding declaration forms two structure variables, *a* and *b*, of structure type complex, which was defined above. Let's say that we have two imaginary numbers, $a = 1.0 + 1.0j$ and $b = 2.95 - 1.0j$. To set our structure variables equal to these numbers, we might use the following assignments:

```
a.real =    1.0;
a.imag =    1.0;
b.real =    2.95;
b.imag = -1.0;
```

The dot notation (*structure.element*) used with the variable names on the left-hand side above allows us to access the individual members of a structure. It is important to understand that the *structure definition* does not allocate a variable but rather simply defines the elements of a structure. We can combine the *structure definition* and the *structure variable* declarations as follows:

```
struct complex {
    float real, imag;
}a, b;
```

This will create the two *structure variables* a and b as before. The equivalent of the above declaration would be

```
float a_real, a_imag, b_real, b_imag;
```

From this, you should see that a structure is just a representation for grouping variables that yields a compact and ordered way to form descriptions of collections of similar elements across different instances of variables. We cannot perform operations on entire structures but only on their individual members using the dot notation described above. In other words, if *a*, *b*, and *c* are structure variables, we cannot form the expression

$$c = a * b;$$

The preceding expression is ambiguous, particularly if the structure contains many different types of variables in its definition.

```
/* multiply two complex numbers */
void c_mult(struct complex a,
            struct complex b,
            struct complex *c,)
{
    c->real = (a.real*b.real)-(a.imag*b.imag);
    c->imag = (a.real*b.imag)+(a.imag*b.real);
}
```

Function 6.2 Multiply two complex numbers.

We can define pointers to structures. To do this, we employ the pointer operator (∗) in the structure variable declaration. Using the complex structure, we can define a pointer to a complex structure as follows:

```
struct complex *c_ptr;
```

Now the c_ptr variable is a pointer to a structure of type complex. Pointers to structures are very useful when writing functions that operate on structures. We indicated that multiplication of structures (and other operations) are not permitted, but we can write a function to perform structure operations. As an example, we will write a function that multiplies two complex numbers. The multiplication of two complex numbers is defined as follows:

```
(a + bj)(c + dj) = (ac-bd) + (ad + bc)j
```

A function to multiply complex variables is shown in Function 6.2. Note that a new notation is introduced, the **structure pointer reference** operator (−>). From the function, the statement

```
c->real = (a.real*b.real)-(a.imag*b.imag);
```

tells us that c is a structure pointer and c->real is the structure member "real." The c_mult function expects to be passed as an address to a complex structure (this is the struct complex *c variable in the argument list). Now we will put the c_mult function in the context of a complete program.

Program 6.3 shows the usage of c_mult in a program. Three complex structures are declared: *a*, *b*, and *c*. The real and imaginary parts of *a* and *b* are initialized before the call to c_mult. Also note that the address of *c* is sent to c_mult, thus causing the results of the multiplication to be returned to the calling program.

```
#include <stdio.h>
struct complex {
     float real, imag;
};
main()
{
     struct complex a,b,c;
     extern void c_mult(struct complex a,
                        struct complex b,
                        struct complex *c);
     a.real=1.0;
     a.imag=1.0;
     b.real=1.0;
     b.imag=1.0;
     c_mult(a,b,&c);
     printf("%f%+f\n",c.real,c.imag);
     exit(0);
}
```

Program 6.3 Program that calls `c_mult` to multiply two complex numbers.

REVIEW WORDS

array
indices
null character
pointer
string
string constant
struct
structure
structure definition
structure pointer reference
structure variable

EXERCISES

1. Write a C function to multiply two four- × -four matrices of type float. How can this program be modified to allow various size matrices to be multiplied?

2. Write C functions to perform complex addition, subtraction, multiplication, and division using the `complex` structure discussed in this chapter. Add these functions to the calculator program that you wrote in Chapter 4. You will have to allow the user to specify a complex variable and input the real and imaginary parts seperately.

3. Define two C structures, one to represent rectangular coordinates and one to represent polar coordinates. Rewrite the `rec_to_polar` function to use variables declared using the new structures.

4. Write a function, `polar_to_rec`, that uses the structures defined in Problem 2. Test your function with various angles and magnitudes.

5. Write a program to calculate simple statistics (use the program developed in Chapter 2 as a guide). Use an array to hold the data set input by the user and use loops to calculate the mean and variance of the data. Also compute the standard deviation.

6. Use the program written in Exercise 5 to determine statistics on a set of outputs of the *rand()* library function. Allow the user to specify how many times the function is called (i.e., how many samples are used in the statistics calculations). What can you conclude from this about the **rand()** function?

7 File Operations

Access to files allows a program to exploit the largest memory subsystem available to contemporary machines, that of the hard disk, CDROM, or tape drives. The file, regardless of which type of media that it resides on, can be thought of as a massive array of data. You can read from this data source, write to it, and create new instances and eliminate them from within your programs. Two primary modes of access to files are available: the low-level operations that work with bytes of data, and high-level operations that utilize data streams to store and retrieve the values of variables. We explore both methods in this chapter.

7.1 Low-Level File Operations

A file can be considered a document that has been stored and that will be accessed as a stream of bytes, such as depicted in Figure 7.1. Files are operating system (OS) resources, and requests must be made to the operating system for access to them. The operating system assigns a **filename** to a file, and the specifics of what characters are permitted in the name as well as the length are system-dependent. Typically, filenames have a prefix and suffix, and by now you have probably encountered this in your programming. The prefix describes the contents of the file, such as "myprog," to describe a program that you are writing. The suffix indicates the type of file, such as ".c" for C source or ".for" for Fortran source, and so on.

To access the data in a file already in the file system, we must first open the file. In C, the **open** function is used to do this. This function has the format

```
fd = open("myfile.dat",mode)
```

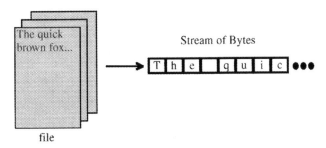

Figure 7.1   File representation.

where

- fd is an integer **file descriptor** returned by the OS,
- *myfile.dat* is either the filename in quotes or a string variable that contains the filename, and
- *mode* is an integer that determines what access will be made to the file, as follows:

mode = 0 → read access
mode = 1 → write access
mode = 2 → read/write access.

If the file does not exist or the user does not have access privilege to the file, the *file descriptor* returned by the OS will be negative. It is a good idea to test the file descriptor to see if the value is nonnegative before trying to read or write to the file.

The **read** function has the format

```
nread = read(fd, buffer, nbytes);
```

where

- fd is the **file descriptor** that was returned by the **open** function,
- *buffer* is an array where the data read will be transferred to,
- *nbytes* is the number of bytes to be read, and
- *nread* is the number of bytes that were read.

The **write** function is similar to the *read* and has the format

```
nwritten = write(fd, buffer, nbytes);
```

where

- fd is the **file descriptor**,

150

- *buffer* is an array of data read to be written,
- *nbytes* is the number of bytes to be written, and
- *nwritten* is the number of bytes that were written.

The file must have been opened for read or read–write, access for read to work and likewise opened to write, or read–write for write to work. If the return variables n_read or or n_written return a negative value, an error has occurred. When the value of n_read is zero, this indicates that there is no more data to be read.

The *open* function assumes that the file exists. If you want to create a new file, you must use the **creat** function. This function has a format similar to *open* as follows:

```
fd = creat("myfile.dat",prot)
```

where

- `fd` is an integer **file descriptor** returned by the OS,
- *myfile.dat* is the file name, in quotes, or a string variable, and
- *prot* is an integer that specifies a protection mode, which is generally expressed as an octal constant.

The modes listed below are for the user:

prot = 0600 → read/write access
prot = 0400 → read only access
prot = 0200 → write only access.

Program 7.1 illustrates the use of all four of these functions. Note that the function prototypes are found in the *stdio.h* header file and the functions themselves are found in the *standard library*. The program gets the filename of an existing file from the user and makes a copy of it to a filename specified by the user. The variables `infile` and `outfile` are the integer file descriptors to be returned by *open* and *creat*. The character array `filename` is used to receive the filename strings inputted by the user to *scanf*. Checks are made of the file descriptors to make sure that they are nonzero. The variable `buf` is a 256-character array that is used to receive the data from reading `infile`. The integer `bufin` is used to determine exactly how many bytes have been read; when `bufin` is zero, the copy is complete. The variable is also used to determine how many bytes to write out to `outfile`.

The last two statements of Program 7.1 introduce the **close** function. The argument to *close* is a file descriptor, and it signals the OS

```
/* File copy program */
#include <stdio.h>

main(){
    int infile, outfile, bufin;
    char filename[20], buf[256];

/* get input filename and open */
    printf("Copy-from filename?");
    scanf("%s",filename);
    infile = open(filename,0);
    if(infile<0){
        printf("File open error...exit!\n");
        exit(0);
    }
/* get output filename and create */
    printf("Copy-to filename?");
    scanf("%s",filename);
    outfile = creat(filename, 0600);
    if(outfile<0){
        printf("File create error..exit!\n");
        exit(0);
    }
/* read up to 256 characters, when bufin = 0,
    stop--otherwise, write bufin bytes out */
    while(bufin=read(infile,buf,256))
        write(outfile,buf,bufin);
    close(infile);
    close(outfile);
}
```

Program 7.1 Low-level file input/output.

to release the file descriptor. When a program terminates, this happens automatically, but use of *close* at the end of the program is good form. There are limits to the number of file descriptors an OS will make available to a program. When a program utilizes many files, file descriptors can be freed up by closing the unused files.

Fortran does not use a low-level file access mechanism. Because the Fortran open, read, and write functions are similar to the C high-level file access functions, we will discuss Fortran file I/O in Section 7.2.

7.2 High-Level File Operations (Streams)

High-level file operations are distinguished from low-level in that the data transfer takes place with buffering and decoding that is not present in the low-level functions. High-level file I/O functions make use of what are called **streams**. A *stream* is just a path for data to take so that it may move from one area of data storage to another. A stream is identified by a **file pointer** as opposed to a *file descriptor*. To access files using streams, a pointer to a stream must be declared as a *file* pointer through the following:

```
FILE *filep;
```

The definition of the typedef FILE is found in the *stdio.h* header file (do not be concerned about what a *typedef* is; for the curious it is a way to rename a data type). We open a *stream* by using the **fopen** function

```
filep = fopen("myfile.dat", mode);
```

where

- `filep` is a **file pointer** returned by the OS,
- *myfile.dat* is the file name, in quotes, or a string variable, and
- *mode* is a string that specifies an access mode.

The most common modes used are listed below:

```
mode = "rw"  →  read/write access
mode = "r"   →  read only access
mode = "w"   →  write only access
mode = "a"   →  append data to file.
```

Unlike the *open* function, if the file does not exist, it will be created. On error, *fopen* will return zero or NULL (defined in stdio.h). One accesses data on an open stream in a variety of ways. The two most common means are with **fprintf** and **fscanf**. These functions are identical to their standard I/O cousins, *printf* and *scanf*, with the exception that the first argument in each case is the stream to be

```
#include <stdio.h>
#include <math.h>

/* program to write a table of square roots
   of x from x=0 to 1.0 at .1 increments to
   a file "xroots.tab" */

main(){
    FILE *fx, *fopen();
    float x = 0;

    fx = fopen("xroots.tab","w");
    if(fx==NULL){
      printf("file open error...exit.\n");
      exit(0);
    }

    while(x <= 1.0){
      fprintf(fx,"%f\t\t%f\n",x,sqrt(x));
      x += 0.1;
    }
    fclose(fx);
}
```

Program 7.2 Use of **fopen, fprintf,** and **fclose.**

read from or written to. The use of these functions is best illustrated by example, because, if you are adept at using *printf* and *scanf,* the transition to *fprintf* and *fscanf* is minor.

Program 7.2 shows a routine that opens a file "xroots.tab" for writing using *fopen.* If the file does not exist, it is created; if it does exist, it is written over. A loop is run to compute the table of square roots. As mentioned, the first argument to *fprintf* is the *file pointer,* in this case *fx.* If we were to print the contents of the "xroots.tab" file after the program runs, it would yield the following:

```
0.000000        0.000000
0.100000        0.316228
0.200000        0.447214
0.300000        0.547723
```

```
0.400000        0.632456
0.500000        0.707107
0.600000        0.774597
0.700000        0.836660
0.800000        0.894427
0.900000        0.948683
```

Hence, the data is in ASCII text format. We "printed" to the file using *fprintf*, just like *printf* "prints" to the console. This is very convenient and allows us to output data that is readily available to other programs. For example, we can output program data to a file for loading to a spreadsheet program for further manipulation or to a math program like Matlab™ or Mathematica™ for plotting.

Program 7.3 shows how the data in the "xroots.tab" file generated by Program 7.2 can be read using *fscanf*. Note that now the file is opened for reading, and as each line is read it is immediately sent to the console using a *printf*. You may want to try each of these programs with your compiler to verify the operation of *fopen, fscanf,* and *fprintf*.

```c
#include <stdio.h>
/* program to output a table of square roots
   of x from x=0 to 1.0 at .1 increments read
   from the file "xroots.tab" */

main(){
  FILE *fx, *fopen();
  float x = 0.1,rootx;

  fx = fopen("xroots.tab","r");
  if(fx==NULL){
     printf("file open error...exit.\n");
     exit(0);
  }

  while(fscanf(fx,"%f\t\t%f\n",&x,&rootx)>0)
     printf("%f\t\t%f\n",x,rootx);
  fclose(fx);
}
```

Program 7.3 Use of **fopen, fscanf** and **fclose**.

Fortran file operations involve the **OPEN, READ, WRITE,** and **CLOSE** intrinsic functions. Fortran file I/O is either formatted or unformatted. If formatted I/O is used, a **FORMAT** statement is needed to define the formatting of the data. We will discuss unformatted I/O only. The Fortran **OPEN** statement has the following syntax:

```
OPEN(UNIT=3, FILE='myfile.dat',STATUS='OLD')
```

You may use any (integer) unit number that you like; however, most operating systems reserve unit number 5 for the standard input (typically the console keyboard) and number 6 for the standard output (typically the console display). The FILE in the open statement must either be a character string or the name of a file in single quotes, as shown above. The STATUS specifies whether you want to open an existing file, STATUS = 'OLD', or create a new one, STATUS = 'NEW'.

The Fortran READ function allows the reading of data from a file. The Fortran **READ** statement has the following syntax:

```
READ(UNIT=3,*)X,Y,Z
```

Here the unit number must correspond to a file that has been opened, and the asterisk (*) indicates that unformatted read will take place. Because Fortran uses variable reference instead of value, no special addressing considerations are needed. In the example above, variables X, Y, and Z will be read. The compiler will take care of any formatting on the basis of variable type.

The Fortran **WRITE** function allows the writing of data to a file. The Fortran WRITE statement has the following syntax:

```
WRITE(UNIT=3,*)X,Y,Z
```

Once again, the unit number must correspond to a file that has been opened, and the asterisk (*) indicates that unformatted write will take place. In the example above, variables X, Y, and Z will be written to the file specified in the OPEN statement for UNIT number 3. The compiler will take care of any formatting based on the type of the variables. If X, Y, and Z are real, decimal points will be outputted.

The Fortran **CLOSE** function is similar to the C *fclose* and *close* functions and has the following syntax:

```
CLOSE(3)
```

PROGRAM FILES

REAL X, Y, Z

OPEN(UNIT=3,FILE='MYFILE.DAT',STATUS='NEW')

X = 3.1426
Y = 1
Z = 2.5E12

WRITE(3,*)X,Y,Z

CLOSE(3)

STOP
END

Program 7.4   Use of Fortran **OPEN**, **WRITE**, and **CLOSE**.

The number in parentheses is the UNIT number opened. Like C, when it is called it frees up the UNIT number for reuse. Program 7.4 illustrates the use of the Fortran OPEN, WRITE, and CLOSE functions. The output of this program (what is written to the file 'MYFILE.DAT') is

```
3.142600    1.000000    2.5000000E+12
```

The compiler has formatted the data automatically with decimal points because they were of Fortran type REAL.

REVIEW WORDS

creat
close
fclose
file descriptor
file pointer
filename
fopen
fprintf
fscanf
open
read

stream

write

EXERCISES

1. Write a C program to process data from a weather monitoring system. The weather monitor provides temperature, barometric pressure, wind speed, wind direction, and humidity information. The data from the monitor can be accessed by a set of system functions: *temp, baro_p, wind_spd, wind_dir,* and *rh* that return their respective values when called. The functions return float values with units and ranges according to Table 7.1. The program should access the data from the monitor; compute the running averages of temperature, wind speed, and relative humidity; determine whether the barometric pressure is rising, falling, or staying steady after three readings; compute the maximum and minimum temperature and humidity; and indicate wind direction using the eight cardinal points of the compass. All of the data should be displayed dynamically to the console at 5-second intervals. If a data value is out of range, a message should be outputted and the program halted. Simulate the weather monitoring system by opening and reading from 5 files containing data for 2 minutes of operation. These files should be named TEMP.DAT, BARO_P.DAT, WIND_SPD.DAT, WIND_DIR.DAT, and RH.DAT.

2. If you wrote a program to compute simple statistics (Exercise 5 from Chapter 6), modify the program to accept data from a file.

Table 7.1 *Weather Monitor Functions.*

Function	Data	Units	Range
temp	Temperature	°F	$-50 \rightarrow 150$
baro_p	Barometric pressure	Inches of Hg	$0 \rightarrow 40$
wind_spd	Wind speed	mph	$0 \rightarrow 120$
wind_dir	Wind direction	Degrees	$0 \rightarrow 360$
rh	Relative humidity	%	$0 \rightarrow 100$

The file can be a simple column of numbers that you can generate using a text editor. Use *fscanf* to determine when the values read from the file are exhausted.

3. Write a program to output values of a loop variable x, the sin (x) and cos (x) to a file. Vary x from zero to 2π. Use a spreadsheet or graphics program to plot the data produced. You may have to separate the columns of data with a comma or tab character depending on what plot program you use.

8 Case Studies

I n this final chapter we examine two programs that have a fair degree of complexity and will serve to bring together many of the concepts of the text. The first program is an adaptation of a simpler program to compute either the height of the tide, given a time of day, or the time that the tide will be at a given height. This type of program finds use in many applications of ocean engineering and is called a *modeling program*, for it models the behavior of a physical phenomenon. Modeling is related to simulation, in which case the computer is used to simulate a process or device. The techniques of computer modeling and simulation are major aspects of engineering practice because they allow us to design and analyze systems in the laboratory before making a high-value commitment to a physical prototype. Modeling and simulation are also useful when working in potentially dangerous and hazardous environments to evaluate risk and develop protective appliances and procedures for personnel who may be exposed to those environments.

The second program shows you how to use a console to plot functions in time. This type of programming is called *visualization* and entails the use of the computer to enhance the presentation of data for analysis or design purposes. Visualization is used in all fields of engineering practice primarily owing to the huge volume of data that engineers must deal with on a day-to-day basis.

8.1 Tides

We first encountered the tides program in Section 5.4 during our discussion of program scope. The program consists of a set of functions that perform different tasks, and we will now discuss each of them in turn while we examine the program source. You may want to refer to the program flow diagram shown in Figure 5.5 at this time. Recall that the program consists of the following functions in addition to the *main*:

```
more        -- asks user to continue or terminate.
TimeOut      -- computes when tide will be out.
TideTime     -- computes time at some height.
TideHeight   -- computes height at some time.
getParams    -- gets almanac data from the user.
TimeFloat    -- converts hours and minutes to minutes.
```

The *main* tides program is listed as Program 8.1. It is somewhat self-documenting. Function prototypes are listed for the four functions that will be called by *main*: these are *more, TimeOut, TideTime,* and *TideHeight*. An event loop formed by a *while* statement that is always true prompts the user to decide whether to compute time or height or to quit the program. A *switch* statement on the character returned by *getch* processes the user request or defaults to an error prompt. The *more* function, listed in Function 8.1, is called to determine whether the user wants to process another data set or to quit. The question is posed, if the user does not want *more* processing then exit, else process data again.

If the user wants to know what time the tide will be at a given height, the function *TimeOut* is called with the function *TideTime* as the argument. *TideTime* is listed as Function 8.2. It takes no argument and queries the user for the desired tide height. It then calls the *get-Params* function (listing Function 8.3) to get the necessary almanac data from the user. This function inputs the time and height of high and low water for the day from the user. These data are available in almanacs, but more sophisticated tide computation programs incorporate a database of this information, and thus the user need not enter the information. Upon return to *TideTime*, the time of the tide at the desired height is computed from an extrapolation formula. This time is in the form of *hour.minute*, a float value where the whole part of the number is the hour and the fractional part the minutes. It is this value that *TideTime* returns to *TimeOut*. The TimeOut function, Function 8.4, converts the float-format time to hours and minutes and displays the results to the user. The integer *hour* is assigned the value of the *floor* of the float value *time*. The *floor* function is in the math library and returns the integer truncation of the variable passed

```
/* Tides
      Program to compute height of tides at some time
              --or--
      time when tide will be at some height.        */
#include <math.h>
#include <stdio.h>

void main(void)
{
    extern unsigned char more(void)
    extern void TimeOut(float TimeIn);
    extern float TideTime(void);
    extern float TideHeight(void);

    while(1){
        printf("\nTide time, height or quit (t/h/q)?")
        switch(getch()){
            case 't':
            case 'T':
                TimeOut(TideTime());
                if(!more())exit(0);
                break;
            case 'h':
            case 'H':
                printf("\nThe tide height will be:
                        %f feet. \n", TideHeight());"
                if(!more())exit(0);
                break;
            case 'q':
            case 'Q':
                exit(0);
            default:
                printf("\nPlease enter a 't', 'h' or
                        'q'!\n");
        }
    }
}
```

Program 8.1 Tides main program.

```
/* Asks user whether further computations are desired,
   returns false if no, true if yes */
unsigned char more(void)
{
    for(;;){
        printf("Compute again(y/n)?");
        switch(getch()){
            case 'y':
            case 'Y':
                return(1);
            case 'N':
            case 'n':
                return(0);
        }
    }
}
```

Function 8.1   Tides *more* function.

```
float TideTime(void){
    float h1, h2, t1, t2, hx;

    printf("\nEnter the tide height that you want the
time for:");
    scanf("%f",&hx);
        getParams(&h1, &h2, &t1, &t2);
        return(
            t2 + (t1-t2)*(acos(1.0 - 2.0*((hx-h2)
/(h1-h2)))/180.0)
        );
}
```

Function 8.2   Tides *TideTime* function.

to it. The *hour* is then subtracted from *time* to yield the fractional part, which is then multiplied by 60 to derive *minute*. The hour and minute when the tide will be at the desired height is then output to the console.

If the user wants the height of the tide at a particular time, then *main* calls the *TideHeight* function, Function 8.5. Note that *TideHeight* takes no arguments but is called as an argument of the *printf* in *main*

```
void getParams(float *H1, float *H2, float *T1, float
                 *T2){
    extern float TimeFloat(int hours, int minutes);
    int hour, minute;

    printf("Enter the Height of High Water:");
    scanf("%f",H1);
    printf("Enter the hour when High Water occurs:");
    scanf("%d", &hour);
    printf("Enter the minute when High Water occurs:");
    scanf("%d", &minute);
    *T1 = TimeFloat(hour,minute);
    printf("Enter the Height of Low Water:");
    scanf("%f",H2);
    printf("Enter the hour when Low Water occurs:");
    scanf("%d", &hour);
    printf("Enter the minute when Low Water occurs:");
    scanf("%d",&minute);
    *T2 = TimeFloat(hour,minute);
}
```

Function 8.3 Tides *getParams* function.

```
void TimeOut(float time){
    int hour, minute;

    hour = floor(time);
    minute = 60*(time - hour);
    printf("The time of the tide will be
%d:%d\n",hour,minute);
}
```

Function 8.4 Tides *TimeOut* function.

that outputs the return value of *TideHeight*, which returns the float value of the computed tide height. *TideHeight* asks the user for the desired time in the form of two integers, *hour* and *minute*. It then invokes *getParams* to get the almanac data. *TideHeight* then returns the computed tide height. Note that the time in hour:minute format is converted to a float format, hour.minute, by the function *TimeFloat*, Function 8.6. TimeFloat casts the hour as float and adds the ratio of the minute to 60 to it.

```
float TideHeight(void){
     int hour, minute;
     float R, h1, h2, t1, t2, tx;

     printf("\nEnter the hour you want to know the
            tide:");
     scanf("%d", &hour);
     printf("Enter the minute you want to know the
            tide:");
     scanf("%d", &minute);
     getParams(&h1, &h2, &t1, &t2);
     R = h1 - h2;
     return(((h1+h2)/2.0)-
     ((R/2.0)*cos(180.0*((TimeFloat(hour,minute)-t2)/
       (t1-t2)))));
}
```

Function 8.5   Tides *TideHeight* function.

```
float TimeFloat(int h, int m){
     return((float)h + m/60.0);
}
```

Function 8.6   Tides *TimeFloat* function.

To test the program, tide data from the newspaper can be used. The following data were published in the *Orlando Sentinel* on May 29, 1997:

```
Daytona Beach, Florida
Units are feet
   Low Tide: Thu 1997-05-29 8:53 PM EDT 0.08
   High Tide: Fri 1997-05-30 3:06 AM EDT 4.10
```

It might be useful to know at what time the tide will be at 2 feet so we run the program and receive the following outputs, user input in **boldface**:

```
Tide time, height or quit (t/h/q)?t

Enter the tide height that you want the time for:2
Enter the Height of High Water:4.10
Enter the hour when High Water occurs:3
```

```
Enter the minute when High Water occurs:06
Enter the Height of Low Water:0.08
Enter the hour when Low Water occurs:20
Enter the minute when Low Water occurs:53
The time of the tide will be 20:43
Compute again(y/n)?y
```

Note that we converted the times to 24-hour format. We can check the result by asking at what height the tide will be at 20:43:

```
Tide time, height or quit (t/h/q)?h
```

```
Enter the hour you want to know the tide:20
Enter the minute you want to know the tide:43
Enter the Height of High Water:4.10
Enter the hour when High Water occurs:3
Enter the minute when High Water occurs:06
Enter the Height of Low Water:0.08
Enter the hour when Low Water occurs:20
Enter the minute when Low Water occurs:53
```

```
The tide height will be:2.322977 feet.
Compute again(y/n)?n
```

There is a slight discrepancy, because of the inaccuracy of the prediction model, but the calculation is certainly close enough for small-craft purposes. It should be clear that the *Tides* program is far from optimal. Many constructs were used to illustrate usage rather than to program either compactly or efficiently. As an exercise, you may want to rewrite the tides program to have fewer calls to functions or possibly eliminate functions completely, such as *more* or *TimeFloat*. An interesting project might be to rewrite the program to access a database rather than query the user for almanac data. These databases can be downloaded from government sites on the World Wide Web. When we computed twice in the same run, the high- and low-tide parameters had to be entered twice. A simple improvement would be to ask the user if he or she wants to enter new tide data and skip the call to *getParams* if the old data are to be reused. A final modification would be to have the program output a plot of the tide height versus time. For this, the second program that we discuss could be useful.

8.2 Console Plot

Plotting involves the graphical display of data as opposed to a tabular display. For example, the data listed in Table 8.1 are far less interesting than the graphic plot shown in Figure 8.1. Tabular data output is easy to produce with just a few lines of code. For example, the data listed in Table 8.1 were produced by Program 8.2. The plot of Figure 8.1 was produced by a computer graphics program, and the program is substantially more complex than that which generated the table. Programming computer graphics is beyond the scope of this book; in fact, entire books have been devoted to it. Nevertheless, the old adage "a picture says a thousand words" holds true for graphic versus tabular data in many engineering applications. For engineers to solve a large spectrum of problems, it is often useful to be able to visualize the behavior of a variable over time or against some other parameter.

We can now pose the question, Is there some way that we could plot sines and cosines using the same output method (i.e., the console) that we use for tabular data? The answer to this question is yes, we can. To explore this concept we need first to understand the layout

Table 8.1 *Sine and Cosine of x.*

x	Sin (x)	Cos (x)
0.00	+0.00	+1.00
0.50	+0.48	+0.88
1.00	+0.84	+0.54
1.50	+1.00	+0.07
2.00	+0.91	−0.42
2.50	+0.60	−0.80
3.00	+0.14	−0.99
3.50	−0.35	−0.94
4.00	−0.76	−0.65
4.50	−0.98	−0.21
5.00	−0.96	+0.28
5.50	−0.71	+0.71
6.00	−0.28	+0.96

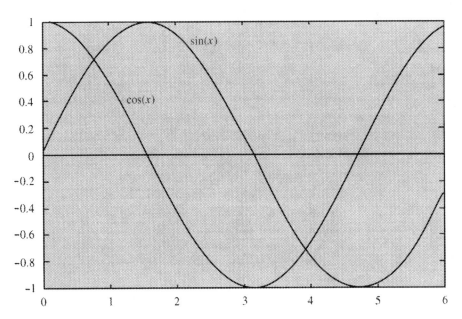

Figure 8.1 Sin (*x*) and cos (*x*) graphics plot.

```
/ * * * * * *Sine and Cosine Table Generator
            This program outputs a table of Sines and Cosines of x
            as x varies from 0 to 2π
* * * * * * /

#include <stdio.h>
#include <stdlib.h>
#include <math.h>

#define PI 3.1416

void main(void)
{
    float x;

    printf ("x\t\t\tSin(x)\t\t\tCos(x)\n");

    for(x=0.0; x<=2*PI; x += 0.5)
        printf("%4.2f\t\t%+4.2f\t\t%+4.2f\n",x,sin(x),  cos(x));

}
```

Program 8.2 Sine and cosine table program.

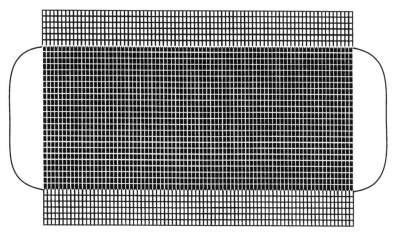

Figure 8.2   Character cells on standard display (80 × 24).

of the standard ASCII display. The size of the display is typically 80 characters wide by 24 lines, and this will be the layout of our "virtual graph paper" for the plotting. Because we can output continuously to the display and it will scroll, we can consider the length of the plot to be infinite. This layout is illustrated in Figure 8.2. We can visualize the display as a window with a continuous sheet of graph paper that scrolls behind it. What we want is to produce a graph of a sinusoid like that shown in Figure 8.3. To do this, we can produce a vertical axis by repeated output of a newline (**\n**), and we can place the plotted value by spacing across each line proportional to the function value.

Choose a plot width of 40 spaces. We know that sines and cosines vary from −1 to 1; therefore, when the function is −1 we don't want to space over at all. When the function is zero, we want to space over 20, and when the function is 1 we want to space over 40. Declare an integer variable P that will be the number of spaces that we need to move over so the following C statement will produce a value for P that is proportional to the spaces needed for the sin function of x:

```
P = (20*sin(x))+20;
```

The plot shown in Figure 8.3 was produced with the following expression:

```
P = (20*sin(w*T + theta))+20;
```

where three new variables are introduced; w for angular frequency; T for time, and *theta* for phase. When plotting with characters on the display it is useful to mark the time axis with the actual value

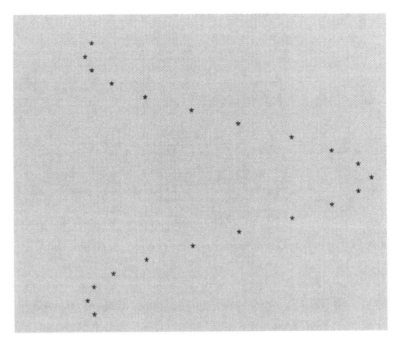

Figure 8.3 Sinusoid plotted on display.

```
>>1  0  0  6.28  .5
1.000000 0.000000 0.000000 6.280000 0.500000
```

```
0.00                          *
0.50                              *
1.00                                 *
1.50                                   *
2.00                                  *
2.50                              *
3.00                   *
3.50              *
4.00        *
4.50  *
5.00  *
5.50     *
6.00            *
```

Figure 8.4 Sin $(T + 0)$ plotted on display with time axis.

of time, and thus we produce a plot, as shown in Figure 8.4. This figure was produced by Program 8.3, which is the prototype program for the plot program we will develop here. This program has a very primitive user interface. The user must know the sequence of data to be entered, shown in **boldface**, and put spaces between the values. Because we are prototyping, the simple interface will be fine while we test ideas for the plotter.

```
/*test program to output sine console plot */

#include <stdio.h>
#include <stdlib.h>
#include <math.h>

void main(void)
{
        float w,theta,Tstart,Tend,dt,T;
        int fcn, P;

    printf(">>");

        scanf("%f %f %f %f %f",&w,&theta,&Tstart,&Tend,&dt);
        printf("%f %f %f %f %f\n\n",w,theta,Tstart,Tend,dt);

      for(T=Tstart; T<Tend;T += dt){
            printf("%5.2fl",T);
            P = (20*sin(w*T + theta))+20;
             while(P−)printf(" ");
            printf("*");
            printf("\n");

      }

}
```

Program 8.3 Sine console plot test program.

We can now summarize what the program does so far as follows:

1. The angular frequency, phase and start, end, and increment value of time are entered.
2. A loop takes the time variable T from $Tstart$ to $Tend$ in steps of dt.
3. On each loop iteration, the time is outputted; then the cursor is spaced by P spaces, where P represents the value of the sine function.

Although the plot looks fairly good, we still have some work to do. It would be useful to add an x-and y-axis to the plot to make it easier to read. The y-axis is implemented by adding the following *printf* before the loop:

```
printf(" t +---------------0-- ------------- + \n");
```

```
>>1 0 0 6.3 .25
1.000000 0.000000 0.000000 6.300000 0.250000
```

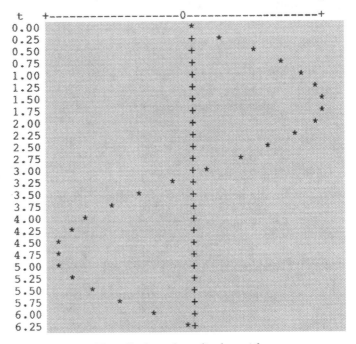

Figure 8.5 Sin $(T + 0)$ plotted on display with axes.

The x-axis, or time axis, is somewhat trickier. We want the output
to look like Figure 8.5, with the axis running down through the center of the plot. The questions are when to plot the axis ('+'), when
to plot the function value ('*'), and when to output a space to move
everything over. One solution is shown in Program 8.4. Here we establish a loop with the integer variable *spc* that keeps track of where
we are on the line as we space across. When *spc* is equivalent to P,
the value of our function in terms of spaces, we output a '*', the plot
marker. When *spc* is equivalent to 20, the center of our plot line, we
output a '+', the axis marker. Note the use of *continue* to make the
loop iterate after a symbol has been outputted. Otherwise, we output
a space to move the plot over. This scheme works regardless of which
order the plot value and axis value are in. It does waste time, however,
in that the output always continues to space over after both symbols
have been plotted. How might you make the routine more efficient?

All that is left to do is to make the interface a bit more friendly,
and this can easily be accomplished by having the program ask for
the plot parameters by name. Another option is to allow the user to

choose other functions for plotting. There are no exercises in this chapter, but you should get Program 8.4 running with the necessary user interface and the ability to plot either sine or cosine.

```
/*test program to output sine console plot
  with x and y axes                    */

#include <stdio.h>
#include <math.h>

void main(void)
{
     float w,theta,Tstart,Tend,dt,T;
     int fcn, P, spc;

   printf(">>");
    scanf("%f %f %f %f %f", &w,&theta,&Tstart,&Tend,&dt);
    printf("%f %f %f %f %f\n\n",w,theta,Tstart,Tend,dt);
         printf(" t +− − − − − − − − − −0− − − − − − − − −−+ \n");
   for(T=Tstart; T<Tend;T +=dt) {
         /*Output the time value  */
         printf("%5.2fl",T);
           /*Compute the spacing value  */
           P = (20*sin(w*T + theta))+20;
           /*Loop for 40 spaces, the maximum width of plot */
           for(spc=0; spc < 40; ++spc){
           if(spc==P) {                    /* plot value */
               printf("*");
               continue;
           }
             else if(spc==20){          /* plot axis */
               printf("+");
               continue;
             }
           printf(" ");                  /* space over */
           }
           printf("\n");                    /* go to next line */
     }
}
```

Program 8.4 Completed prototype sine console plot test program.

Appendix A:
C Language Summary

The summary is alphabetic with reserved words or required symbols in boldface. Definitions for items in <> are listed.

---

arithmetic operators

```
add       +      3 + 4 = 7
subtract  -      3 - 4 = -1
multiply  *      3 * 4 = 12
divide    /      3 / 4 = 0.75
modulus   %      3 % 4 = 3
```

---

arrays

```
array_name[<size>]                        /* 1-D */
array_name[<size>][<size>]                /* 2-D */
array_name[<size>][<size>][<size>]        /* 3-D */
array_name[<size>]···[<size>]             /* etc.*/
```

---

ascii character

A single byte-sized character, equivalent to ASCII binary code. See Appendix C.

```
'A'  ⇔  41₁₆
```

---

ascii string

A set of ASCII characters.

```
"Hello"
```

---

assignment

$x = 3;$

Can be preceded by an arithmetic or bitwise operator:

```
x += 1;    ⟺ x = x + 1;
x <<= 3;   ⟺ x = x << 3;
```

---
bitwise operators

```
and           &    1 & 0 = 0
or            |    1 | 0 = 1
not           ^       ^1  = 0
shift left    <<
shift right   >>
```

---
break

Unconditional exit from loop.

---
case

See **switch**.

---
cast

Converts variable or result of expression to type.

```
(<type>) <variable or expression>
     (int)7.0/3.0 ⟺ 2
```

---
character constant

```
'<ascii character>'
          c = 'a';  ⟺ c = 0x61;
```

---
character string constant

```
"<ascii string>"
    c = "This is a string.";
```

---
comment

```
/* this is a comment in C */
```

<compound statement>

All statements within the braces will be treated as a single statement.

```
{
    <statement>
        ⋮
    <statement>
}
```

---

continue

Causes loop to go to next iteration of control variable.

---

decrement operator

```
--X; ⇔ X = X - 1;
```

---

default

See **switch**.

---

do-while loop

Executes statement as long as the test <expression> evaluates true.
Do executes first then tests.

```
do <statement> while( <expression>);
```

---

engineering (scientific) notation

```
3.0E9     ⇔   3,000,000,000.0
3.0E+9    ⇔   3,000,000,000.0
2.07E-4   ⇔   0.000207
```

Exponent must be integer.

---

escape characters

```
\n - newline    \t - tab
\f - formfeed   \" - quote
\b - backspace  \r - return   \\ - backslash
```

<expression>

Can be a number, variable, arithmetic or logical operator combination, or null. See <statement>.

```
3.1416
X
x + y
x && y
x & y
i
++j
x = y - z
```

---
for loop

A controlled loop with a start, stop, and increment definition.

```
for(<expression 1>;<expression 2>;<expression 3>)
     <statement or compound statement>
```

Note:

Expression 1 is evaluated at the start of the loop.

Expression 2 is evaluated to determine if the loop should continue (true: continue, false: stop).

Expression 3 is evaluated at the end of each iteration through the loop.

---
function call

```
<label>(<parameter values>);
```

examples:

```
get_some_value(value);
exit();
```

---
function definition

```
<type> <label>(<parameter list>){<statements>}
```

example:

```
int get_some_value(float *value){
     return(fscanf("%f",value));
}
```

---
goto

Unconditional transfer of execution to statement following <label>.

```
goto <label>;
```

---
if

If <expression> evaluates true, then statement or compound statement is executed; otherwise, it is skipped.

```
if(<expression>)<statement or compound statement>
```

---
if-else

```
if(<expression>)<statement or compound statement 1 >
     else <statement or compound statement 2 >
```

If `<expression>` evaluates true, then `<statement or compound statement 1 >` is executed; otherwise `<statement or compound statement 2 >` is executed.

---

increment operator

```
++X;  ⇔  X = X + 1;
```

`<label>`

Any sequence of alphabetic and numeric characters, including the underscore, that starts with a letter. The number of allowable characters depends on the compiler.

---

logical operators

Logical operators cause an expression to be evaluated to a Boolean true (1) or false (0) result. They are used in conditional and loop expressions (if, for, switch, do, etc.).

```
AND                 &&
OR                  ||
NOT                 !
```

`<parameter list>`

A sequence of `<type>` `<variable>` combinations separated by commas that establish the parameter types to be passed to a function. The 'double *x*, double exponent' in the parentheses below constitute the *parameter list* for the pow function.

```
double pow (double x, double exponent);
```

`<parameter values>`

A sequence of variable names or constant values to be passed as the parameters to a function. `'y'` and `'2.3'` are the *parameter values* for the function call to `'pow'` below:

```
x = pow(y,2.3);
```

In C, if a variable is passed as a parameter value, the function receives the *value* of the variable.

`<NULL statement>`

The null statement consists of a single semicolon. No execution takes place. Can be used to form infinite loops.

```
;
```

---

relational operators

Used to establish logical relations between variables or constants. In the examples, all of the statements formed using the operators are true.

```
equivalence                 ==   5 ==5
not equal                   !=   5 !=3
less than                   <    5 < 6
greater than                >    5 > 3
less than or equal to       <=   5 <=5
greater than or equal to    >=   5 >=3
```

---

return

Statement used to return from a function call execution. If no return statement is used, the function returns null.

```
return;        /* return from function */
return(x);     /* return value of x */
return(x<2);   /* return logical result */
```

---

sizeof

```
C function that returns the size in bytes of a
variable.
              sizeof(<variable>);
```

<size>

```
Integer variable or integer constant.
See arrays.
```

<statement>

```
A statement is an <expression> followed by a
semicolon.
```

---

switch

A **switch** allows multiple conditions to be evaluated against a single expression. The <expression> is evaluated and compared against the variable or contant in each **case**. If they match, the **case** statement(s) is (are) executed. If a **break** statement is placed with the case, execution begins at the first statement following the **switch**; otherwise, the next **case** is evaluated. The **default case** is always executed if execution did not break under previous cases.

```
switch(<expression>){
    case <variable or constant>:
        <statement(s)>
        break;
            ⋮
    case <variable or constant>:
        <statement(s)>
        break;
    default:
            <statement>
}
```

<type>

Type declares the data type of the variable or expression. The size in bytes of a type is dependent on the host machine architecture.

int	integers
short int	integers(smaller range)
unsigned int	positive integers
long int	integers(larger range)
float	reals(rationals)
double	reals(larger range)
char	ASCII characters
void	null type

---

typedef

Allows redefinition of a type for clarity in program style.

```
typedef <type> <label>;
```

<variable>

A variable is a data storage element defined by a label and of a specific type. A variable must be declared at the beginning of the program or externally.

```
<type> <label>;
```

---

while loop

Executes <statement>as long as <expression> tests true. **While** <u>tests first</u> and then executes if true.

```
while(<expression>)<statement>;
```

Appendix B: Fortran Program Language Summary

This summary is alphabetic with required symbols or words in boldface. Definitions for items in <> are listed.

---

arithmetic operators

```
add              +        3 + 4 = 7
subtract         -        3 - 4 = -1
multiply         *        3 * 4 = 12
divide           /        3 / 4 = 0.75
exponentiation   **       2 ** 4 = 16
```

---

arithmetic IF

```
IF(<arithmetic expression>)<n>,<z>,<p>
```

If the result of the expression is negative, program execution transfers to statement <n>; if zero, to statement <z> and to statement <p> if positive.

---

arrays

```
DIMENSION array_name (<size>)
DIMENSION array_name (<size>,<size>)
DIMENSION array_name (<size>,<size>,<size>)
```

1-D, 2-D, and 3-D arrays. Note that <size> may be C different integer values or an integer variable.

---

ascii character

A single byte-sized character, equivalent to ASCII binary code. See Appendix C.

---

ascii string

A set of ASCII characters.

---

assignment

$x = 3$

CALL subroutine

Used to invoke a subroutine, for example,

```
CALL MYSUB(X,y)
```

---

character constant

```
'<ascii character>'
c = 'a'
```

---

character string constant

```
'<ascii string>
c = 'This is a string.'
```

---

column positions

```
1      -- Letter 'C' or '*' in column
indicates comment line.

1-5    -- statement number.

6      -- any character in this
column indicates continuation.

7-72 --  FORTRAN statement.

73-80 --  ignored by compiler; used to
number statements for
           programmer reference or for
           comments at end of line.
```

---

comment

```
C this is a comment in Fortran
                ('C' in first column)
```

```
* this is also a comment in Fortran
                    ('*' in first column)
```

CONTINUE

Causes loop to go to next iteration; used as terminator for a loop.

DO loop

```
DO <statement> <index> = <start>,<end>,<increment>
```

Executes all statements up to <statement> from <index> equal to <start> until <index> is greater than the <end> value. On each iteration of the loop, the value of index is incremented by <increment>. If no increment value is given, <increment> defaults to 1.

END

```
END of program.
```

engineering (scientific) notation

```
3.0E9 3.0E+9 2.07E-4  Exponent must be integer.
```

<expression>

Can be a number, variable, arithmetic or logical operator combination, or null. Examples:

```
3.1416  x + y  i  x = y - z
```

<field list>

A list of field definitions used in a **FORMAT** statement. A field definition consists of an optional count (allows more than one variable to share the same field definition), a format code, and a code format. For example, a FORMAT statement with a list of field definitions is as follows:

```
99 FORMAT(1X,F20.2,3I10)X,I,J,K
```

The list contains one X code, one F code, and three I codes. The F and I codes correspond to the variables X, I, J, and K.

```
FORTRAN Format Codes

F   --    real number
D   --    double precision real
E   --    scientific notation
I   --    integer
L   --    logical
X   --    blank (space)
H   --    alphanumeric character
```

For real variables, the format code is written as one of the following: cF$w.d$ c, D$w.d$ or cE$w.d$ where c is the repeat count, w is the width of the number in character positions, including the decimal point and sign, and d is the number of characters after the decimal point. All other codes are simply followed by the character width required.

FORMAT statement

Used to define the size and type of data for I/O. Format statements must be numbered.

```
<statement number>FORMAT(<field list>) <variable list>
```

function call

```
<name>(<parameter values>)
```

FUNCTION definition

```
<type> FUNCTION <label>(<parameter list>)
        <statements>
  RETURN
```

GOTO

Causes execution to transfer to <statement number>.

```
GOTO <statement number>
```

IF

```
IF(<logical expression>)<statement>
```

If expression evaluates true, then statement is executed.

IF-THEN-ELSE

```
IF(<logical expression>)THEN
        <statement(s)>
    ELSE
        <statement(s)>
    ENDIF
```

If expression evaluates true, then statement(s) is(are) executed; otherwise, the statement(s) following the else is(are) executed.

Any sequence of the characters below starting with a letter. Number of allowable characters is typically six.

```
ABCDEFGHIJKLMNOPQRSTUVWXYZ
1234567890
```

---
library functions

See table at end of this appendix.

---
logical operators

AND	.AND.
OR	.OR.
NOT	.NOT.
EQUIVALENT	.EQV.
NOT EQUIVALENT	.NEQV.

Truth Table for Logical Operators

A	B	.NOT.A	A.AND.B	A.OR.B	A.EQV.B	A.NEQV.B
F	F	T	F	F	T	F
F	T	T	F	T	F	T
T	F	F	F	T	F	T
T	T	F	T	T	T	F

A sequence of <variable> combinations separated by commas that establish the parameter types to be passed to a function. An example is as follows:

```
CART_TO_POLAR(x,y,mag,angle)
```

<parameter values>

A sequence of variable names or constant values to be passed as the parameters to a function. An example is as follows:

```
CART_TO_POLAR(2.3,4.5,mag,angle)
```

relational operators

equal to	**.EQ.**
not equal	**.NE.**
less than	**.LT.**
greater than	**.GT.**
less than or equal to	**.LE.**
greater than or equal to	**.GE.**

RETURN

Return from function or subroutine.

```
RETURN
```

<size>

Integer variable or integer constant.

<statement number>

Integer value between 1 and 99999.

STOP

STOP execution.

SUBROUTINE definition

```
SUBROUTINE <label>(<parameter list>)
      <statement(s)>
RETURN
```

<type>

Type declares the data type of the variable or expression. The size in bytes of a type is dependent on the host machine architecture.

Appendix B: Fortran Program Language Summary

INTEGER	integers
REAL	reals(rationals)
DOUBLE PRECISION	reals(larger range)
CHARACTER	ASCII characters
LOGICAL	TRUE or FALSE

<variable>

<type> <label>

The following functions are common to Fortran compilers:

function	operation performed
ABS	real absolute value
ALOG	natural log
ALOG10	base 10 (common) log
AMOD	real modulus
ARCOS	arc cosine
ARSIN	arcsine
ATAN	arctangent
CONJ	complex conjugate
COS	cosine
COSH	hyperbolic cosine
EXP	exponential, ex
FIX	convert real->integer
FLOAT	convert integer->real
IABS	integer absolute value
MOD	integer modulus
SIN	sine
SINH	hyperbolic sine
SQRT	square root
TAN	tangent
TANH	hyperbolic tangent

Appendix C: ASCII Tables

Code in Hexadecimal – ASCII Character

00	NUL	01	SOH	02	STX	03	ETX	04	EOT	05	ENQ	06	ACK	07	BEL
08	BS	09	HT	0A	NL	0B	VT	0C	NP	0D	CR	0E	SO	0F	SI
10	DLE	11	DC1	12	DC2	13	DC3	14	DC4	15	NAK	16	SYN	17	ETB
18	CAN	19	EM	1A	SUB	1B	ESC	1C	FS	1D	GS	1E	RS	1F	US
20	SP	21	!	22	"	23	#	24	$	25	%	26	&	27	'
28	(	29	)	2A	*	2B	+	2C	,	2D	–	2E	.	2F	/
30	0	31	1	32	2	33	3	34	4	35	5	36	6	37	7
38	8	39	9	3A	:	3B	;	3C	<	3D	=	3E	>	3F	?
40	@	41	A	42	B	43	C	44	D	45	E	46	F	47	G
48	H	49	I	4A	J	4B	K	4C	L	4D	M	4E	N	4F	O
50	P	51	Q	52	R	53	S	54	T	55	U	56	V	57	W
58	X	59	Y	5A	Z	5B	[	5C	\	5D	]	5E	^	5F	_
60	`	61	a	62	b	63	c	64	d	65	e	66	f	67	g
68	h	69	i	6A	j	6B	k	6C	l	6D	m	6E	n	6F	o
70	p	71	q	72	r	73	s	74	t	75	u	76	v	77	w
78	x	79	y	7A	z	7B	{	7C	|	7D	}	7E	~	7F	DEL

Appendix C: ASCII Tables

Code in Decimal – ASCII Character

```
0    NUL  1    SOH  2    STX  3    ETX  4    EOT  5    ENQ  6    ACK  7    BEL
8    BS   9    HT   10   NL   11   VT   12   NP   13   CR   14   SO   15   SI
16   DLE  17   DC1  18   DC2  19   DC3  20   DC4  21   NAK  22   SYN  23   ETB
24   CAN  25   EM   26   SUB  27   ESC  28   FS   29   GS   30   RS   31   US
32   SP   33   !    34   "    35   #    36   $    37   %    38   &    39   '
40   (    41   )    42   *    43   +    44   ,    45   -    46   .    47   /
48   0    49   1    50   2    51   3    52   4    53   5    54   6    55   7
56   8    57   9    58   :    59   ;    60   <    61   =    62   >    63   ?
64   @    65   A    66   B    67   C    68   D    69   E    70   F    71   G
72   H    73   I    74   J    75   K    76   L    77   M    78   N    79   O
80   P    81   Q    82   R    83   S    84   T    85   U    86   V    87   W
88   X    89   Y    90   Z    91   [    92   \    93   ]    94   ^    95   _
96   `    97   a    98   b    99   c    100  d    101  e    102  f    103  g
104  h    105  i    106  j    107  k    108  l    109  m    110  n    111  o
112  p    113  q    114  r    115  s    116  t    117  u    118  v    119  w
120  x    121  y    122  z    123  {    124  |    125  }    126  ~    127  DEL
```

Code in Octal – ASCII Character

```
000 NUL 001 SOH 002 STX 003 ETX 004 EOT 005 ENQ 006 ACK 007 BEL
010 BS  011 HT  012 NL  013 VT  014 NP  015 CR  016 SO  017 SI
020 DLE 021 DC1 022 DC2 023 DC3 024 DC4 025 NAK 026 SYN 027 ETB
030 CAN 031 EM  032 SUB 033 ESC 034 FS  035 GS  036 RS  037 US
040 SP  041 !   042 "   043 #   044 $   045 %   046 &   047 '
050 (   051 )   052 *   053 +   054 ,   055 -   056 .   057 /
060 0   061 1   062 2   063 3   064 4   065 5   066 6   067 7
070 8   071 9   072 :   073 ;   074 <   075 =   076 >   077 ?
100 @   101 A   102 B   103 C   104 D   105 E   106 F   107 G
110 H   111 I   112 J   113 K   114 L   115 M   116 N   117 O
120 P   121 Q   122 R   123 S   124 T   125 U   126 V   127 W
130 X   131 Y   132 Z   133 [   134 \   135 ]   136 ^   137 _
140 `   141 a   142 b   143 c   144 d   145 e   146 f   147 g
150 h   151 i   152 j   153 k   154 l   155 m   156 n   157 o
160 p   161 q   162 r   163 s   164 t   165 u   166 v   167 w
170 x   171 y   172 z   173 {   174 |   175 }   176 ~   177 DEL
```

Useful escape codes:

NUL -- null	NL -- newline
BEL -- bell or beep	VT -- vertical tab
BS -- backspace	NP -- newpage
HT -- horizontal tab	CR -- carriage return
ESC -- escape	DEL -- delete

Appendix D:
C Preprocessor
Directives

The C Preprocessor is a filtering mechanism that is applied to a C program source file prior to compilation. The preprocessor may be a front end to the compiler or a separate program that is called by the compiler. The purpose of the preprocessor is to allow the programmer to control the compilation at the source level. The three primary uses of the preprocessor are

- to allow inclusion of header or definitions files (so that sources common to multiple programs can be easily managed and reused),
- to allow definitions of special constants or macros to minimize typing and simplify program structure, and
- to allow partitioning of the source and control which portions are compiled.

Commands, or directives, to the preprocessor are prefixed with a pound sign (#). For example, the most common of the preprocessor directives is the *include* directive:

```
#include <stdlib.h>
```

Directives are terminated by a newline. If a directive requires more than one line, then newlines must be escaped for each line except the last. For example, the following *define* macro requires more than one line:

backslashes to
escape newlines

```
#define quartic_on_x(x) \
    pow(x,4) + a1*pow(x,3) + a2*pow(x,2) \
    + a3*x + a4;
```

This appendix discusses the three most common commands to the preprocessor: the include, the define, and the conditional. Boldface type is used to show the directive and its required parts, whereas normal type is used for the user-specified portions.

#include

The include directive allows you to specify the name of a file (stored by the operating system) for inclusion in your source code prior to compilation. When the preprocessor encounters an *include* directive, it searches for the filename specified. If the filename is located, the preprocessor inserts the text of the file into the source at the location of the directive. Remember that this takes place prior to compilation. If the file is not located, the preprocessor halts (as does compilation) with an error message.

The *include* directive has two primary forms:

```
#include <stdlib.h>
#include "myheader.h"
```

The first form (< >) causes the preprocessor to search for the filename specified, in this case *stdlib.h*, in a set of user-specified directories and then in a standard set of system directories. This form should be used when including system header files. Examples of these types of files are *stdio.h, conio.h, float.h,* and so on.

The second form is used when you wish to include a header file of your own that is specific to your program. The quotes (" ") instruct the preprocessor to search for the file in the current directory, the directory containing the source being compiled. The only difference between the two is determined by where the files are located. The rule of thumb is to use the first form for system headers and the second for user files. The *include* process is illustrated by the graphic below:

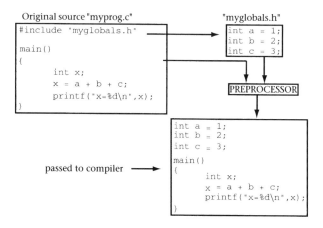

A third form exists for the *include* directive that allows you to compute the file argument as a macro. The explanation for how to use this form is beyond the scope of this text.

#define

The *define* directive allows for user-specified definitions and macros. A definition is used when you want the preprocessor to replace a label that you define with something else. The label is generally uppercase to distinguish it from program variables, but this is not required by the directive. A few examples follow:

#define PI 3.1415927

This is a commonly used definition. Whenever the label *PI* appears in the source, the preprocessor will replace it with the constant 3.1415927. In this example, PI is not a variable. A usage such as this relieves you of repeatedly entering the constant, makes the program easier to read, and allows you to control the interpretation (fix precision and accuracy) of a constant.

#define OK printf("OK so far...\n");

Use this when you want to avoid typing a lengthy expression many times throughout the program. In this example, whenever *OK* appears in the program, the preprocessor replaces it with the *printf* expression. This is useful when trying to debug and track program execution dynamically.

Define is also used to create *macros*. A macro is a compact representation of a more complex expression. By this definition, the simple *define* expressions above are also macros, but typically a macro contains replacement variables. When the *define* label contains parentheses, a macro is specified. Here is an example of a macro:

```
#define CIRCUMFERENCE(r) 2.0*PI*r
```

replacement variable

The example assumes that PI has been defined already (see above). If CIRCUMFERENCE(5.0) is encountered in the source, the preprocessor will replace it with $2.0*3.1415927*5.0$. You may also use multiple variables:

`#define` SECTOR_AREA(r,theta) 0.5*r*r*theta

replacement variables

Conditional Compilation

Conditional compilation directives allow you to control which portions of the source are compiled. This is useful in many instances, but the two most popular are when a program is to be "crippled"* for distribution of a demonstration version and to allow for dynamic debugging codes to be removed prior to final compilation or to remove portions of code during debugging. The directives discussed here have the following syntax:

> `#ifdef` <*defined symbol*>
>   *program statements*
> `#endif`

or

> `#ifndef` <*defined symbol*>
>   *program statements*
> `#endif`

The first case asks whether or not the < *defined symbol* > has been defined. If it has, then the program statements between the **#ifdef** and **#endif** lines are compiled; otherwise, they are not passed on to the compiler. The second case is the logical inverse and asks if the symbol has *not* been defined.

In the example below, the *printf* statements will be included in the source passed to the compiler because the symbol *DEBUG* is defined. Each run of the program will output the value of *x* at those points of execution allowing the variable to be tracked. When program execution is satisfactory, the *#define DEBUG* can be removed for final compilation, and the *printf* statements will be removed.

---

*A crippled program has program features disabled.

```
#define DEBUG
main(){
              ⋮
        <program statements>
              ⋮
 #ifdef DEBUG
      printf(">value of x is:%f\n",x);
 #endif
              ⋮
        <program statements>
              ⋮
 #ifdef DEBUG
      printf(">value of x is:%f\n",x);
 #endif
              ⋮
        <program statements>
              ⋮
}
```

Appendix E: Precedence Tables

C Precedence Rules

Operator	Associativity
() [] -> .	$\Rightarrow$
! ~ ++ -- (type) * & sizeof	$\Leftarrow$
* / %	$\Rightarrow$
+ -	$\Rightarrow$
<< >>	$\Rightarrow$
< <= < >=	$\Rightarrow$
== !=	$\Rightarrow$
&	$\Rightarrow$
^	$\Rightarrow$
|	$\Rightarrow$
&&	$\Rightarrow$
||	$\Rightarrow$
?:	$\Leftarrow$
= += -= *= /= %=	$\Leftarrow$

In Fortran, all expressions associate left-to-right ($\Rightarrow$) except exponentiation, which evaluates right-to-left ($\Leftarrow$).

Fortran Precedence Rules

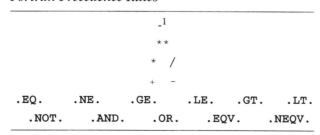

```
                    _1
                   **
                 *   /
                 +   -
 .EQ.    .NE.    .GE.    .LE.    .GT.    .LT.
   .NOT.    .AND.    .OR.    .EQV.    .NEQV.
```

[1]Unary.

Glossary

ADA® – Highly structured, object-oriented language developed by the Department of Defense and named after the first software programmer, Ada Lovelace.

address – binary word that determines the location of a cell of data in memory.

algorithm – a detailed set of instructions on how to perform a calculation.

analytical engine – Babbage's second calculating machine. It was designed to be programmable using punched cards.

application – executable program that performs a useful task.

arithmetic logic unit (ALU) – the part of a computer architecture that performs arithmetic and logical operations on data.

arithmetic operator – operator that acts on variables or constants numerically.

array – a collection of same-type variables with a distinct access scheme based on indices for members in the collection.

ASCII – acronym for American Standard Code for Information Interchange and the 7-bit binary code used to represent symbols.

assembler – application that converts assembly language files into object code.

assembly language – computer language that is unique to a processor class and is constructed of mnemonic codes that describe transfer of data between registers and other portions of a computer architecture.

bells and whistles – capabilities added to a program or application that go beyond the basic requirements of the task to improve the user interface.

binary – base 2 numbering system. Numbering system with radix 2.

bit – binary digit, the fundamental unit of information storage in a computer.

bitwise operator – operator that acts on variables or constants as Boolean values.

byte – a binary word that is 8 bits wide.

C – A high-level, problem-oriented language originally developed to write operating systems but now used for any programming task requiring speed, efficiency, and portability.

cascaded errors – errors caused as a result of prior error. Failure to form or terminate a statement properly is often a cause of cascaded error.

cast – explicit means within C to perform forced type conversion, for example,

$$x = (\text{float})\, i + z;$$

casts variable i to float prior to addition with variable z.

cell – storage unit in a main memory system.

compile-time error – refers to errors that the compiler flags. Syntax, type mismatch, and macro errors are examples of compile-time errors.

compiler – application that interprets high-level language code into object code for linking and loading.

conditional – a statement, such as an *if* statement, that controls execution based on a condition.

control unit – the part of a computer architecture that orchestrates passage of data and the operations performed by a computer architecture.

debug – remove the errors from a computer program.

decrement – to decrease the value of a variable, generally by 1.

development environment – specialized program that integrates an editor, compiler, and run-time facility.

difference engine – Babbage's first calculating machine. It was designed to compute navigation tables using a method of finite differences.

I–O – input–output.

dummy variables – variable names used in a function or subroutine definition that are simply place holders and have no stored values.

edit–compile–run cycle – term given to the process of preparing an application in a high-level language, the programming cycle.

editor – program used to enter source code into a computer or to create ASCII document files.

entry point – place in a program or function where execution starts.

explicit – not implied, requires direct statement or definition.

file – a set of binary data that a computer uses to store and manage information.

flag variable – a variable used to signal a condition.

flowchart – graphical technique of diagramming a program.

Fortran – FORmula TRANslation. A high-level, problem-oriented language designed for scientific and engineering programming.

function – a program within a program.

function prototype – a C declaration that defines the type of a function and the types of its pass variables.

global – term signifying that the scope of a variable or function is available to any part of a program.

implicit – implied, not directly stated.

increment – to increase the value of a variable, generally by 1.

index – number, symbol, or expression used to indicate position in a loop, array, or sequence.

infinite loop – a loop that has no loop control variable and thus no predefined stopping point.

intrinsic function – functions that are built in to a particular language; C has only one, *sizeof( )*.

label – name given to a variable or program element.

library function – collections of functions available to the programmer whose source code is not part of the program.

linker–loader – operating system application that combines object code with libraries to create applications.

local – term signifying that the scope of a variable is restricted to the defining function.

logical operator – operator that acts on variables logically (with values of true or false) as opposed to arithmetically.

loop – a program structure that allows a program to double back and execute the same statements multiple times.

loop control variable – a variable that is tested to signal the end of a loop; variable that controls the number of times a loop executes.

machine code – binary codes that are interpreted as instructions by the hardware of a computer.

mass storage unit – large I–O unit used to store files, most often a disk drive.

matrix – multidimensional array.

nibble/nybble – a binary word that is 4 bits wide.

object code – assembled instructions. Files containing object files are an intermediate step in the development of an application.

off-line – operation performed without the computer.

operating system – special application used to manage the resources of a computer system.

operator – a functional rule that is indicated by a symbol or set of symbols.

operator precedence – ordering rules that determine when an operator acts in an expression.

pass by reference – phrase denoting when the address of the pass variables in a subroutine or function is passed (Fortran and Pascal use this method).

pass by value – phrase denoting the value of the pass variables in a subroutine or function are passed (C and ADA use this method).

program – a sequence of instructions describing how to perform a task.

promotion – when a variable of one type is converted to that of a higher type (e.g., float to double).

pseudocode – literally means "false code." A simple, easily understood, written description of a program. Not intended to be compiled.

radix – base of a numbering system.

register – the smallest unit of memory in a processor; used to hold data for memory, ALU, or I/O operations.

run-time error – refers to errors that occur at run time or after an application has been compiled successfully. Divide-by-zero is a common run-time error.

scope – the range of availability of a variable or function within a program.

source code – a computer program in higher-level language form, such as a C or Fortran program.

string – an array of characters.

syntax error – an error that violates a language structure or format rule.

text only – file or data in ASCII printable code.

type – classification of variables, as in data type.

type conversion – implicit and explicit rules and procedures for converting from one type to another.

variable – a defined and labeled data storage element in a program.

vector – a single-dimensional array.

virtual machine hierarchy – a structure that partitions a computer system into higher levels of abstraction from the hardware level.

von Neumann machine – computer architecture with program storage in memory.

word – basic unit of memory or register size in a computer system (expressed as a collection of bits).

Annotated Bibliography

C Programming

A. Feuer, *The C Puzzle Book*, Prentice Hall, 1982.

> Clever and well-presented collection of C programming tidbits and puzzles. An excellent source of material for honing skills in C programming.

K. Jamsa, *The C Library*, McGraw-Hill, 1985.

> Large collection of well-written and commented C code for string manipulation, pointers, array manipulation, recursion, sorting, and file operations.

A. Kelley and I. Pohl, *Turbo C: The Essentials of C Programming*, Addison-Wesley, 1988.

> Good treatment of Borland Turbo C for the beginning programmer.

A. Kelley and I. Pohl, *C: by Dissection*, Benjamin/Cummings, 1987.

> Excellent tutorial text for the intermediate-level C programmer.

B. W. Kernighan and D. M. Ritchie, *The C Programming Language* (2nd Ed.), Prentice Hall, 1988.

> Recognized classic text as the original definition of the C programming language.

W. H. Press, B. P. Flannery, S. A. Teukolsky, and W. T. Vetterling, *Numerical Recipes in C*, Cambridge University Press, 1988.

> Recognized classic compendium of scientific C functions and programs. Diskette available with code.

R. Sedgewick, *Algorithms in C*, Addison-Wesley, 1990.

> Comprehensive collection of complex algorithms implemented in C.

Fortran Programming

D. M. Etter, *Structured FORTRAN 77 for Engineers and Scientists* (4th

Ed.), Benjamin/Cummings, 1993.

Popular textbook for Fortran 77 engineering and scientific programming.

C. Lampton, *FORTRAN for Beginners*, Franklin Watts, 1984.

Good starter book for basic FORTRAN programming.

J. F. Kerrigan, *From Fortran to C*, Windcrest, 1991.

Written for FORTRAN programmers to learn C. A very good dual reference and tool for advanced study of both languages.

Index

Index

Index

Index

Printed in the United States
By Bookmasters